P9-DFW-089

GENETIC FIX

The Next Technological Revolution

AMITAI ETZIONI

HARPER COLOPHON BOOKS
Harper & Row, Publishers
New York, Hagerstown, San Francisco, London

For Mike and Dari, my youngest

the text of this book is printed
on 100% recycled paper

This book was first published in 1973 by
Macmillan Publishing Co., Inc.
It is here reprinted by arrangement.

First HARPER COLOPHON edition published 1975

ISBN: 0-06-090428-3

78 79 80 81 82 10 9 8 7 6 5 4 3

Contents

Preface

If you are concerned with one or more of the following questions, you had better inform yourself about the new genetics, your chances, your rights, and your opportunities.

—Could you, and should you, be tested to find out if, unbeknownst to you, you are carrying in your genes a disease which will send you into a wheelchair later in life? (Researchers have identified more than fifteen hundred illnesses which are partially or completely determined by our genes.)[1]

—If you plan to have another child, do you know the steps you can take *now* to help it to be normal and to avoid having a retarded child caused by phenylketonuria (PKU), or a mongoloid (a tragedy which hits one out of every six hundred births in the United States), or a victim of at least a score of other devastating afflictions?

—If you want your next child to be a boy or a girl, do you know the steps necessary to achieve the desired result? And how to arrange for some other biological features, such as height, complexion, and perhaps IQ?

—If your kidneys or some other vital organ fail, do you know what will determine whether or not a replacement can be found? Whether the cash needed will be available? Whether the doctors and nurses and technicians needed will have been trained in the proper techniques?

—As a citizen, should you approve, disregard, or oppose experiments to grow babies in test tubes and to make xerox copies of people? (Experiments in these areas are now in progress.)

—Should our government demand that everybody's genes be checked before they get married? Should people with "criminal" chromosomes (XYY) be forbidden to have children?

—Should the new genetics be used to breed a healthier and superior race?

—And who should make all these decisions? You? Your doctor? The government? A council of wise persons?

These and other such questions must be faced in the near future because yet another technological revolution is upon us. We are witnessing the close of an era, an era shaped by the Industrial Revolution, dominated by economic and technocratic concerns, driven by a gigantic productive machinery. It has also been an age in which the increase in our capacity to think, to analyze, and, above all, to provide humane, responsible direction to our efforts did not keep pace with the growth of our muscle power. We became a sort of mighty Frankenstein creature, sometimes quite mindless and rather heartless.

The new revolution, based on developments in both biology and physiology, will do to our genes and brain chemistry what the Industrial Revolution did to our muscle power. Natural processes will become far more subject to human engineering. In this new age, "givens" will be transformed into issues we can debate, fight, and decide about. The new revolution will vastly expand our capability both to wreak evil and to render good; it will thrust upon us awesome new powers and responsibilities, from the capacity to mass-produce identical infants to the ability to stamp out congenital diseases.

True, at the moment we are only at the beginning phase; we cannot yet grow babies in a lab to a designer's specifications; the day we can leave an identical copy of ourself behind may be decades away; and correcting a defective gene or composing a new one, may well be much further away. But there are a great many things we can do now, this year, or in the very near future. I choose to focus on these developments, both to give the discussion a greater measure of reality than the deliberations about distant futures, and because the personal, social, and moral issues raised by the steps we can take now are basically the same as those that future developments will raise. To highlight that we are in the initial phase of the age of genetic *engineering*, I refer to a genetic "fix"; we can *now* tamper with our biological inheritance, both fix it where it went wrong and improve it. Moreover, while for a time we avoided active interventions in the field—they reminded us of the Nazis' eugenic attempts—there is now an increasingly growing demand for a genetic elixir, a genetic fix.

An alternate title for this volume, which I feared would detract from its seriousness, is *What Your Doctor Did Not Tell You, and What Your Congressman Does Not Know, About the New Genetics.* For this book is not primarily a chronicle of new scientific developments; rather, it focuses on your right to know of opportunities and their consequences, your right to be in on the decisions now being thrust upon us, your right to decide if you wish to administer the genetic fix to yourself, your next child, and—with other fellow citizens—to the nation's biological inheritance and future.

This volume is a part of my effort to focus attention on the profound implications of genetic engineering—for you, for me, for our society, and for the future of the human race. What will be done to us next, in the name of science and progress? And what can we do to channel the new force in the direction we want it to go in order to improve the quality of our life?

I see a need for a double antidote: a stimulant to energize the passive members of the community, who allow science to shape society to suit whatever developments the test tubes happen to yield, and a mild tranquilizer for those who hysterically reject all technological advances. What we need most is a greater ability to discriminate, to judge, and to review—in short, a greater maturity. We need these abilities so that we can benefit from the new, powerfully revolutionary scientific breakthroughs without being overwhelmed by them. We must learn to use the new instruments to our benefit, rather than submit to exploitation by them. The basic question is: *Will we use these dramatic innovations to reestablish the priority of our values, or will we simply let them join other instrumental forces which drag us blindly toward their own ends or are abused by tyrannical rulers?*

This somewhat personal volume deals chiefly with what I learned and felt when I participated in an international meeting of experts that reviewed the new scientific breakthroughs, the foundations of tomorrow's technology. It also deals with what I have tried to do about all this and how far I have gotten (not very far, let me tell you right away).

Why a personal account? Because simply to cite only what the experts have stated, would contribute to the separation of mind and emotion that is the curse of modern civilization. If I were merely to list the number of people who die because kidney machines are not available or because a nationwide tissue-match system has not been organized, or tell of the number of infants who have passed away in horrible convulsions because of the experiments conducted on them (see pages 25–26, 137–141), or give statistics on the thousands of unwanted mongoloid babies, I would be giving short shrift to the agony, personal tragedies and societal cost caused by the untrammeled, abused, or unused new technology. Hence, while relevant data are reported in this book, and while sources are given where documentation can be indicated (in the Notes, beginning on page 251, for each chapter), I felt compelled also to attend to the human element behind the figures.

Similarly, the abstract terms of sociology, psychology, or political science cannot fully express the problems raised by the new advances in biological and medical research. These terms do not quite capture the difficulties encountered by myself and others who are trying to elicit from scientists, doctors, lawmakers, moralists, and the general citizenry the commitment that is essential if we are to critically assess and wisely use scientific progress. Hence, while an explicit sociopolitical theory underlies both my analysis and my recommendations,[2] I hope my report of what happened, when I and others tried to move in on these issues, may offer a sense of the complexities of the issues involved and the forces at work.

If more people, who give a damn, will inform themselves about these issues, they in turn will be able to educate the general public and help it act on these matters—matters too vital to be left to the experts or to any government.

Unlike several other treatments of the subject, this book focuses on the *near* future. The year 2000 is too easy to predict—few of us will be held accountable if we prove to be false prophets; and anyhow, there is rather little we can do about a time so remote. But several vital questions—both personal and public—must be answered and acted upon in the immediate future, and it is on those questions that this volume focuses.

Acknowledgments

The chief sources for this volume are the papers presented and the comments made by the participants in a conference organized by Dr. Simon Btesh, the executive secretary of the Council for International Organizations of Medical Sciences. Papers presented at the conference are quoted or summarized here, but the full text is not included. Comments made by the scientists at the conference are given as made, with little editorial change; but, again, only selected comments and not the full exchanges are reported. In few instances comments made on one day have been cited in another, when the same topic reappeared, to reduce redundancy. Copies of the full proceedings are available from CIOMS (c/o World Health Organization, Avenue Appia, 1211 Geneva 27, Switzerland). The participants and organizers of this conference are all co-authors of this volume. I am indebted to them for permission to quote them.

I am indebted for comments on previous drafts of this volume to Sue Bruser, Nancy Castleman, Joshua Freeman, Ani Hurwitz, Carol Morrow, and Mary Helen Shortridge. Ms. Morrow also served as a diligent, thoughtful, and conscientious research assistant throughout this project. Stephanie Fins helped in proofreading. I am grateful for comments on previous drafts of the manuscript to Prof. Bernard Barber and to Prof. James Wechsler, and for discussion of "Are We Debasing Our Genes?" with Prof. Arthur G. Steinberg and Prof. George Fraser. Leonard Castleman, M.D., was most generous in answering several medical queries. Sue Bruser, Pat Fleury, and Lynn Friedman retyped several versions of the manuscript. Gwen Cravens edited it most thoroughly.

The Center for Policy Research made this work possible.

Introduction: A Trip to Paris

When I recently received an invitation to an international conference to be held in Paris, I scarcely hesitated. These days, professors are always being asked to jet to Kyoto or Dubrovnik or Lake Belaggio to share their expertise, though often these junkets are not all they're cracked up to be. But the thought of Paris in September was especially seductive. Tourists gone, Parisians back, little restaurants reopening ...

But it was not just Paris in the early fall and one more chance for professoring that made me accept the invitation with alacrity and set aside my fear of flying and of other travails of travel. The topic of this particular conference was to be "Recent Progress in Biology and Medicine: Its Social and Ethical Implications." A timely topic indeed; developments in both fields are accelerating and their consequences, for better or worse, for health and death, are rapidly multiplying.

Such conclaves are often called by a group or an institute with a warm heart—and a patron—but with little experience in conducting an international conference. Frequently an international set of hacks, operators, and *apparachiks* turn up; the few unworldly scholars or genuine experts who come, get lost in the shuffle, and the meetings soon degenerate into a gossipy muddle of cocktail parties and sightseeing excursions. But the Paris conference was being held by the Council for International Organizations of Medical Sciences (known as CIOMS), a creation of the World Health Organization and UNESCO. With such a sponsor, this conference could not help but be of value.

The issues raised by new developments in biology had already affected me in a rather personal way. In 1967 I was the father of three fine boys, and my wife and I were contemplating having another child. Could we make it a girl? Somewhere I had read a report that the sex of an embryo could be determined by the stage of the menstrual cycle at which fertilization takes place. After a talk with a doctor and a day's trip to the medical library, I learned more about the question of sex choice than I ever cared to know. While I found out there was no choice open to us at the time (and our fourth boy joined us soon thereafter), I did discover that scientists were working on various sex-choice techniques. I also came across some surprisingly reliable data which suggested that if everyone were given the same freedom of choice I was seeking, their greater happiness would result in fairly serious dislocation for the general public. Studies showed that if each set of parents got what they wanted, a considerable "male surplus" would result. The social consequences of such a surplus, I concluded, ranged from the unfavorable to the undesirable. There would be a growing proportion of the population unable to find marriageable partners; homosexuality and prostitution would probably increase. Furthermore, since there would be fewer women, there would be fewer persons interested in culture (books, theater, art) or charged with the moral upbringing of children (which is still a woman's specialization); and there would be more people engaged in competitive, materialistic pursuits (still, more of the man's world). For the same reason, violent crimes would rise (90.4 percent of all violent crimes and 81.3 percent of all crimes against property committed in the United States in 1970 were committed by men).[1]

I'm always embroiled in one public cause or another, and so I did not hesitate to share my newly found wisdom with the readers of *Science* (for a reprint of the article, see Appendix 7), and—with the help of the popular press—with the readers of newspapers around the world which picked up the story. (You can hardly miss their attention when your topic is sex.) Over a period of time, my interest expanded, and I went on to study and write about the social and moral questions raised by other recent and anticipated breakthroughs in biological and medical research. Then came the invitation. Onward to Paris!

Part I:
THE FIRST DAY

CHAPTER ONE

An Overview: What Are They Brewing for Us?

As I sat on the dais in Room X of UNESCO House in Paris, about to deliver one of two introductory presentations to the "round table" of assembled experts, I felt anxious. The hall was gigantic. In addition to the front rows, occupied by scientists from near and far, there were rows upon rows of seats, each with a desk and microphone, reserved for ambassadors of the UNESCO countries, other rows for representatives of international scientific associations, and still others for the press. Booths for simultaneous translation overlooked the hall.

Dr. Alfred Gellhorn, Dean of the School of Medicine of the University of Pennsylvania, presided. He raised the key issue of the meeting in his opening remarks. It was no longer being taken for granted, he pointed out, that progress in medical and biological research was a good thing. He cited eight recent reports all debating "the adequacy of society's wisdom in dealing with the explosive increase in biomedical knowledge." He continued: "Just as we recognize that nuclear energy may be used for constructive or destructive purposes, so the spectacular finding that DNA is the chemical basis for heredity so increases man's knowledge of nature as to lay on him a tremendous responsibility to use wisely this newly acquired power through knowledge."

I tried to read the faces opposite me in the hall: did they wonder, as I did, what reason we had to expect man to be wiser in this area than in any other? A line came to me which I frequently use in lectures and which always elicits a roar of approval: "If there is anything foolish which can be done, sooner or later there will be a government that will do it."

Gellhorn was now listing some of what he called "the nightmare possibilities of the basic scientific discoveries." He referred to the possibilities of cloning—producing a host of genetically identical persons, that is, xeroxing *people,* the result of which "might be dull *at best.*" He said that the same procedures of genetic engineering that are now being developed to correct those genetic deficiencies causing severe illnesses might also be used to breed people of unusual physical strength.

Dr. Gellhorn did not weigh the enormous burden the availability of such a procedure would impose on society. What to breed? Football players, foot soldiers, or musicians? Energetic, combative people, or low-key, pacific, compliant ones? A refined breed that would do well only in an ultra-affluent environment, or a rugged race able to survive a breakdown of modern civilization?

The near impossibility of reaching a consensus on what breed to cultivate, and hence the deeply divisive effect this would have on a society that might have to face such decisions in a future, which might be uncomfortably close, was illustrated by two side remarks I picked up during the conference. One was on the light side; the other, quite serious. Both references were to the work of the late Dr. H. J. Muller, a distinguished professor at Indiana University and a leading advocate of genetic engineering. Muller wrote that a sperm bank of *superior* genes—from, say, geniuses and baseball heroes—should be set up for people wanting artificial insemination. A French professor remarked, re Muller: "What is superior? It would never have occurred to me to include baseball players; I would have insisted on bicyclists." Another professor easily topped this remark: "In an early edition, Muller favored including Stalin among the political donors of sperm for his bank. The name was dropped in later ones." (Actually, it was Lenin who was so cavalierly dropped, but the point was well made.) Not only was one man's hero another man's knave, but—for the same person—yesterday's saints were today's gods who failed. Woe to the woman who used last year's fashionable sperm!

"The authorities in the field have reservations about the practicality of these genetic black magics," Gellhorn continued. It seemed to me, however, that even before genetic engineering developed very far, the mere question of how it might be used would invite a resurgence of racist ideologies and conflicting racist camps, each advocating its version of the desired breed. There are troublesome indications that society is already heading in this direction anyhow. From the thirties until quite recently, in reaction to Nazi racist theories and in response to the struggle of blacks and other minorities for social equality, biological factors have been played down and the role of education and equality of opportunities have been stressed. In recent years, though, with the conservative backlash, the question of whether certain races are inherently inferior has been resurrected.

The first major breakdown in the public's and political leadership's confidence in the educatability of all men and women followed the 1966 publication of a report by a distinguished social scientist, James S. Coleman.[1] The most surprising finding of the study he directed was that, on the average—based on a variety of objective measurements of school quality, including buildings, equipment, class size, teachers' qualifications, curriculums offered, and many others—no significant differences were found between the schools attended by various racial and ethnic groups. The study's main conclusion was that there is no correlation between a child's progress (measured by verbal as well as non-verbal achievement tests) and variation in school facilities, staff, and equipment. In other words, the inequalities that are imposed on children come from their homes, their neighborhoods, and their peer environments; schooling is not the source of the difference, nor is it a greater equalizer. Hence, it is not that important; actually, surprisingly unimportant. Therefore, giving more education to disadvantaged children will not make any difference by itself. The report was subject to a great many interpretations; but the one which gained the most currency seems to be a new pessimism about education, and more indirectly, a greater openness toward considering racial differences as having a measure of permanency, and—a genetic base.

Arthur R. Jensen and Richard J. Herrnstein subsequently published articles arguing that intelligence was largely determined by genetic factors.[2] (Jensen claims that 80 percent of the variance is due to genes; only 20 percent to environmental factors.) In the same years, several reports stated that a variety of mental illnesses heretofore attributed to social or psychic factors were caused by genetic or physiological factors.[3] All this was crowned and ideologically amplified by systematic attacks against the concept of equality by neo-conservative intellectuals such as Nathan Glazer and Irving Kristol.[4]

Thus the ground was readied for the next crop, without anyone consciously plowing in that direction: let us improve the genes to improve society. A change, then, in social climate, and a technological development are once more egging each other on. But is this the right direction? Is the newfound popularity of biologistic interpretation anything more than a reaction, maybe an overreaction, to the previous excessive reliance on education and social reforms as a propellant of change and to the view of man as highly pliable? And will we not, in one or two decades from now, overreact again to the overreaction, after many millions of people may have been affected by what looks to be the next fad, biological engineering? Is there any way to avoid, or at least to reduce, this gross "oversteering," these "enthusiasms," as bases for public policy, and to gain a better grasp of the potentials and dangers involved before we change course?

Another thought crossed my mind as Gellhorn spoke: many viewed biological engineering in the same way they viewed other scientific advances—as one more giant step in the development of modernity toward ever-greater opportunities and ever more freedom of choice in an ever more modern world. But this development, it seemed to me, would, by requiring conscious selection of what has heretofore occurred naturally—should your next child be blond and blue-eyed, or dark; highly charged or low-keyed; tall or short; and so on—almost surely overwhelm the human capacity, already severely taxed, to make sensible choices. Moral disorientation and ad-hoc codes have even now replaced to a great extent the set taboos and traditions of earlier generations. Transitory social relations (e.g., living together) and frequent divorces undermine the stability of the family. Social and geographic mobility weaken community bonds. It all amounts to what Warren G. Bennis and Phillip E. Slater[5] called "the temporary society," in which nothing can be taken for granted any more; all is in flux. The result is not just an expansion of our options and freedoms, but also a proliferation of anxieties and frustrations. Even before we have learned to cope with the endless dynamism of our society and culture, we now face efforts to "temporize" our bodies. Soon our descendants' biological composition and features will no longer be "given" by nature but will be subject to our decisions, fads, and anxieties.

Fortunately, the supermen assembly lines I was reflecting upon seemed to be a long way off. Gellhorn turned to more immediate problems: "It is now *possible* to detect the completely normal, heterozygous carriers of genes [having a pair of genes which determine a particular trait, which are not identical] producing recessive disorders and further of detecting the affected state in a fetus at an early age." (It was suggested later, by someone from the floor, that anyone considering marriage ought to submit to an inexpensive, easily administered genetic test.)

I probably would not have known what Gellhorn was talking about if I had not recently learned how amniocentesis works. Doctors, by drawing out and examining a few drops of the fluid in which a fetus floats, can determine whether or not it will be born a mongoloid.*

Gellhorn was saying that we can already "detect sixty genetically caused defects, and in a few years we should be able to detect over one hundred. The list of deformities which could thus be prevented, by aborting the fetus, include, aside from mongolism, which occurs surprisingly often, chromosome abnormalities associated with Klinefelter's syndrome [small testes, usually no living sperm, high likelihood of mental retardation; occurs in about one of every four hundred baby boys], Turner's syndrome [female sex organs remain undeveloped, webbed neck, dwarfism, puffy hands and feet, bowlegged; heart abnormalities are common; occurs in one out of every two thousand to three thousand baby girls], and the XYY chromosome formation."

*One gynecologist described amniocentesis as follows: The doctor "guides a four-inch needle through the belly wall, into the peritoneal cavity, through the uterine wall, and, lastly, into the amniotic sac. All this must be done without nicking a blood vessel or any of the blood-filled sinuses that are laced around the uterus. Once you get the needle inside the sac, it mustn't penetrate the fetus itself or any portion of the umbilical cord, which may be looped in any position. You don't push the needle in blindly, however, trusting to luck. You must know how the baby is lying, you locate the placenta so you can avoid it, and you push that needle in with the greatest caution, testing as you go. Still, it's a great relief when you get a clear tap—amniotic fluid and no blood."⁶

Only in a few of these, Gellhorn pointed out—and the deliberation of the next days surely bore him out—was the decision about what to do relatively easy. Mongoloid children are prone to acute leukemia and often die young; if they survive, most are doomed to life in public institutions, because most parents cannot cope with them emotionally or economically and tend to abandon them there. The agony for parents, the deformed children, and the costs to the public (1.7 billion dollars a year in the United States alone) make it *relatively* easy to determine the advised course—abortion. But the choice of action concerning other defects is less clear-cut. Gellhorn cited the XYY chromosomal abnormality, which "has been discovered in a significant proportion of violent criminals and in inmates of institutions for the criminally insane. *But* neither do all persons with the particular antisocial behavior have these chromosome changes nor do all those with XYY configuration *conform* to the criminal criteria."[7]

I had read that studies about "the criminal genes" were less reliable than most.[8] The finding has not been replicated sufficiently often to be considered adequately verified, and the data are based on the examination of small groups of subjects, without use of control groups for comparative purposes. It is quite possible that XYY genes are as common in all or some nonincarcerated populations as among inmates. Nor is it known how common these genes are among inmates and to what extent they "predispose" behavior—that is, do they make people slightly more aggressive or deeply affect their whole personalities? But even if future studies would lend more support to the finding, should parents be told when the mother carries a "potentially criminal" fetus? And, I wondered, would a day soon come when society would pressure or require, such parents to abort their "criminal" unborn children?

Gellhorn explained that several other genetic illnesses, like phenylketonuria (PKU), are now treated effectively through dietary therapy. In the past, PKU caused mental retardation and a shortened life-span; therefore afflicted persons seldom bore children to carry on the gene. Now the metabolic disturbances are prevented so that a normal life-span is achieved. But in a sense even this is a mixed blessing because, as a result, people who carry these defective genes are now more likely to reproduce, thereby increasing the number of carriers of defective genes. As other incapacitating genetic illnesses can similarly be corrected, the time may come when every other person may carry one sick gene or another, if not several. And the more the members of such a population marry among themselves, the more genetic sickness will be produced in their offspring.

It occurred to me that one solution might be to put all future afflicted children on the same life-saving diet presently given the smaller number, but detection of the illness, which must be made very early in life for treatment to be effective, is quite difficult.

There are reports of "false positives"[9] (one source placed the figure at 85 percent in 1963), which incorrectly identified children as sick, and led to their being put on a diet that was very restrictive and damaging to a healthy child.[10] A number of misdiagnosed children have suffered from physical deterioration as a result of the diet, while some children are reported to have died. Though the diet is safer today, there is still a danger of malnutrition and subsequent intellectual impairment, as well as behavioral problems and loss of fine motor functions.

In addition, there are reports of "false negatives"—that is, of many PKUs who pass undetected and go untreated—expected, according to one source in 1966, to run as high as 53 percent.[11] Since then, the requirements for carrying out the test have been greatly relaxed and the test, it is believed, improved. Still, one must also consider the matter more broadly: as more and more genetically malformed people are helped by advances in medicine to reach the reproductive stage, will we not see an ever-more illness-prone population? Will we reach a stage where our facilities will be so overburdened that society will be unable to carry more?

Gellhorn, who by now had the rapt attention of his audience, paused briefly before he turned to yet another aspect of the complex issue before the conference—organ transplants. At this point, he explained, these transplants have not become widespread largely on account of the gaps in our immunological knowledge. If we could better match donors to graftees, or deal otherwise with the rejection processes, far more transplants would be made, with hundreds of thousands of people benefiting from them each year.

Yet it is not only a question of budgeting large funds to run such a transplant program, for, as a recent newspaper report pointed out, thousands were dying each year not only because there were not enough dialysis machines for kidney patients but also because we lacked nurses and other personnel trained in their use.[12] Moreover, since the machines have provided a less than complete cure, transplants will have to be increased. In 1971–1972, although an estimated fifteen thousand transplants were needed, only about three thousand were performed because there were not enough donors.[13] And once the habit of leaving one's kidneys to the living catches on, the country is expected to be short of the necessary surgical teams.

Above all, Gellhorn went on, there will continue to be a severe shortage of donors and organs. Who, then, will be allocated the scarce spare parts? Who, following what criteria, will render these you-will-live, he-will-die judgments?

The matter, it seems to me, is much more difficult than the decision to turn off a life-maintaining machine that artificially prolongs the existence of a terminally ill, comatose body, though even this decision is one that many doctors find very disturbing. Decisions about who shall be allotted an organ and thus a very good chance to lead a normal life, and who shall be refused and condemned to die, are as agonizing as those that have to be made on a sinking ship when the lifeboats are too few to carry all the survivors. However, the medical decisions are more tragic, because the mortality rate is quite predictable, and given a greater amount of resources, sufficient life-saving procedures could be prepared.

What an absurd situation! When transplants in general were rare, the allocation problem was manageable; most patients neither received nor expected them. But now, as transplants, especially of kidneys, have become available to thousands of persons, other thousands are being denied them, and the scope of the dilemma is reaching terrible dimensions. What criteria shall we use? Women and children first? Will the rich and the powerful again get the lion's share? Who will be entrusted to stand in for God?

Gellhorn called attention to the fact that the same basic procedures that may help cure still another set of illnesses may also lead to grand-scale manipulation of human behavior. One example is provided by dopamine, which removes the symptoms of Parkinson's disease, a disabling disorder of movement. Other closely-related chemicals affect other specific brain-controlled behavior patterns such as fear, pain, and appetite. While such new drugs may lead to the pharmacologic therapy of major psychoses, the same kinds of drugs may also be employed to make people easy to manipulate.

Gellhorn cited evidence suggesting that these were not very serious prospects, but I was thinking about Dr. Kenneth B. Clark who, while serving as president of the American Psychological Association, proposed the creation of new drugs to be routinely administered—especially to leaders holding great power—to subdue hostility and aggression and to "allow more humane behavior to emerge." Full implementation of his notion is unlikely, but drugs are already being used to control pupils and mental patients. In many schools, hyperactive children are given amphetamines (Ritalin and Dexedrine) that stimulate a normal adult nervous system but slow down hyperactive children. Hyperactive children are unable to sit still and learn without medication, but—significantly—teachers are also giving these drugs to normal children, who are highly active and difficult to handle, in order to make them more manageable.[14] Mental patients, especially those in state mental hospitals, are frequently tranquilized into docility so that aides can be free to do other work and patients will refrain from complaining about abuses, which range from unnecessary use of force to rape.[15] During the 1959 international meeting of the Planned Parenthood Federation in New Delhi, a leading Indian scientist, Dr. Homi Jehangir Bhabha, noted that economic progress for India required a 30 percent reduction in fertility and asked "if we might put something in the food which would reduce fertility." The Indian government did not act on his suggestion, possibly because of the unavailability (so far) of such a substance; the moral issues that the use of such a drug would raise may all too quickly be disregarded when the fear of over-population is dominant. Is it really so far-fetched to think a totalitarian government might use drugs to tranquilize whole populations?

For example, Nobel Prize winner Salvador Luria, an MIT biologist, "has described the possibility of devising a genetic message that would make human beings sensitive to a simple gas like carbon dioxide—in effect, making it a deadly poison. If the gene then could be hooked to a virus and spread throughout a nation, such a weapon would be much more insidious than a bomb; one nation could tamper with the genes of another without its ever knowing it."[16]

Thus, no one could confidently deny the possible arrival of a bio-chemical 1984 before the seventies are out.

By now, Gellhorn was expressing his "personal conclusion": The way to secure beneficial effects instead of "deleterious applications" was through a dedication to ethical values, to "enlargement of individual dignity, full opportunity for human potential, the realization of human spirit." Had it been someone else speaking I may have wondered whether he had chosen to express these fine sentiments and Sunday platitudes because such an invocation is the accepted way for the presiding officer to open this sort of convention. But I felt sure that Gellhorn, fine M.D. that he is, would rather rely on the "human spirit" than on government authorities to sort out these matters. Most doctors oppose any regulation of "their" business.

My thoughts were interrupted when I heard our moderator introducing me. I was on next.

Should We, Could We, Edit Science?

I began by reacting to Gellhorn's lofty closing note:

"I'm interested in the mechanisms and the institutions necessary to guide science and technology," I said. "I would maintain that values rarely fly on their own. They need social institutions to protect and sustain them. My specialty and concern is the condition under which normative choices can be implemented."

Most participants of scientific meetings write papers beforehand and then proceed to read them aloud. Gellhorn had read his in carefully balanced sentences and an even voice. I had found on past occasions that when I read a paper it sounded as if I were reading a text which someone else had written and which I did not quite comprehend. So I usually ad-lib, with an occasional glance at my notes. This time was no exception.

"My interest is in the particular condition under which we can control our fate, but to avoid a misunderstanding, let me define my general position," I continued, drawing both on what I had written on the subject earlier and on what I had just heard. "It's so easy in these discussions to paint the other side into one of two corners—to see a person as either in favor of stopping progress, curbing science, destroying technology, and returning to the Stone Age; or in favor of a complete free reign for science, preferably with no questions asked. If any control has to be introduced, this second group thinks it should be done by the individual practitioner and perhaps by his peers. My position is simply a middle one. I favor control, but featherweight. I think some questions must be asked not by each doctor on his own, when the spirit moves him, but by a duly instituted body.

"I'll go into details of my position in a moment but let me first cite an example. Subliminal advertising flashes a message on a television screen too rapidly for conscious perception. Nevertheless, the eye transmits the message to the brain. Experimenters have tried to use subliminal advertising to convince people to buy popcorn. But we can easily imagine messages of a less innocent nature being conveyed.

"So far, all experiments with this technique have been unsuccessful. Yet, some of my colleagues are still working on subliminal communications. [17] And if they have a good night's work tonight, you will read tomorrow morning on the front pages of your newspaper that it is now possible to control people through their television screens without their consent and without their awareness. Another case in point is the analysis of the E Waves in standard EEG tests. These allow doctors to identify the person's sexual thoughts, e.g., does he feel stimulated by a male or female?[18] It is thus now possible to detect if persons have homosexual tendencies, even if they do not wish to admit to them, and without their knowledge that they are being tested. The question is, as long as sexual deviance is punished, as long as people may lose their jobs and be subject to cruel stigmas, should such a procedure be developed?

"The question these examples raise is not whether we should stop science or do away with modern technology. We couldn't if we wanted to. The question is: *Can we edit progress?* What mechanisms, used in what ways, make it possible to curb some undesirable effects without hindering the mainstream of development?"

I reminded the participants that the procedure used in determining genetic defects in a fetus can now also be used to learn the sex of the embryo, and I asked whether we should leave only to the discretion of each physician the decision about whether amniocentesis should be used for this purpose. "We must take into account that if sex choice is practiced on a large scale, a serious sex imbalance would be created, at least in the United States. According to my calculations about seven percent more males than girls would be ordered each year (see Appendix 7). This conclusion is based on data about the sex composition of families who practice birth control. We find more people will stop having children after they have had only boy babies or boy and girl babies than if they have had only girls (see Appendix 7). Now I want to survive the cocktail party, so let me add that I did not *cause* this finding. It is not my fault that people so choose; I'm just reporting it. The study also shows that the resulting sex imbalance would strain every social institution we know. One of the first to be hurt—and its already smarting enough —is the family.

"Now I don't wish to turn this into an evaluation of sex choice; I just wish to highlight the question and get to my major thesis. In the past we have not only talked, but acted, as if society must absorb, with little say in the matter, much of what science uncovers and what can be achieved technically. Whether it was nuclear weapons, steam engines, or whatever, the unlimited freedom of science meant that society had to adapt. If the inventions of science and technology meant that people had to congregate into cities, millions of people moved from the countryside. Society had to give way and find some way to accommodate itself. Modern society stands largely on the primacy of science and technology over other social institutions.

"My thesis to you is that at this point we do not have, even on paper, the mechanisms for stopping a particular development once it is proven undesirable. For instance, some researchers stumbled upon something known as LSD. Next, we are admitting many thousands of people to hospitals for psychotic episodes because chemists put some things together and somebody in the Harvard Department of Psychology tried them in an experiment; from there LSD spread right into the bloodstream of hundreds of thousands of young Americans. And there is nothing we can do if I report to you that right this minute somebody is composing some other chemicals more dangerous than LSD. We might unanimously agree that it should not be, but there is, so far, no mechanism we have now to stop it."

I went on to point out that if the community of scientists and physicians procrastinates until biological and behavior-modifying breakthroughs in chemistry ambush us in some kind of disaster, then the politicians and the government will curb them. An overreaction, a censorship on all research, might certainly come about. This is a far from apocryphal statement. Much research in the USSR is guided, precisely for this reason; and during the Middle Ages, the Catholic Church did its best to control research. Research on abortion or amniocentesis is still very restricted in Catholic nations. In Israel, where religion opposes the use of cadavers, pathologists run into repeated difficulties. Other countries have other restrictions. In short, the freedom of science cannot be taken for granted.

"It is a question, therefore, of finding that golden mean whereby we will have opportunity for 'editing' progress under the auspices of a commission composed of members of the theological, humanist, medical, and scientific communities, so that the curbing can be done with the gentleness and sensitivity, which I believe the disciplines call for," I said.

I spent the rest of the twenty-five minutes allotted to me discussing the more general sociological conditions under which societal processes can be guided rather than allowed to meander. I had previously tried, in an overly long book, to outline what I considered the sociopolitical analogue of the Keynesian theory, which tells roughly where the levers are for the guiding economic processes. Ever since *The Active Society* was published, I've tried to summarize its main points in almost every public presentation I've delivered; I consider it my life's work, and try to make up for the difficulties that its length seems to pose.[19] This time around, the subject was the conditions under which we can deal with societal problems caused by science rather than being overwhelmed by them. In this regard, I felt the wisest strategy of attack would be the establishment of a commission to deal specifically with the social and ethical issues raised by genetic interventions and new breakthroughs in medicine—in short, the establishment of what might, for the sake of brevity, be referred to as a Health-Ethics Commission.

The economists, I reminded my audience, have developed a theory which tells us what we have to do if we wish to regulate the economic process instead of being subject to its whims. We now need such a theory for societal processes, since, as in economics, our muscle power has grown more rapidly than our capacity to reflect and guide, and unless we catch up in our regulatory capacity, we will have to expect more—and more damaging—mindless societal zig-zags.

A Keynesian theory of societal processes, to put it briefly, suggests we need at least four elements to form an effective societal regulatory capacity: (1) *Knowledge* of the processes involved (until recently we had only little fragmental information on the ups and downs of societal processes and the forces which propel them); (2) a "middling" *decision-making strategy*, which is neither as demanding and impractical as total planning nor as incremental and myopic as the current "muddling-through" approach; (3) authentic *consensus* on goals and views, and new ways of reaching *consensus* more rapidly, as more decisions must be made, as the societal business mounts; (4) finally, a distribution of *societal power* supportive of *broad-based* change rather than concentration of power in the hands of a few elites, bureaucrats, and technocrats.

This done, I rested, and tried not to show that I enjoyed the rather warm and fairly prolonged applause I received. The chairman said "Thank you, Professor Etzioni, for a very stimulating paper," and sounded as if he meant it.

Later, during a coffee break, people told me how much they liked my paper. I had not anticipated such a favorable response. A portly, drab-suited young man whose tag said "Dr. Fraser" congratulated me. The one theologian at the conference, Prof. Jürgen Moltmann, said something approving in a heavy German accent. Mrs. Gellhorn was particularly flattering, and Dr. Gellhorn added, "You and I are not as far apart as it seems," a remark with which I hastened to agree. When Prof. Maurice Lamy of the French L'Académie Nationale de Médecine suggested that I join him for lunch. I was moved. Prof. David Klein of the Institut de Génétique Médicale at the University of Geneva came along with us. When he learned that I was not Italian but Jewish (people often think my name is Italian rather than Israeli), he congratulated me in fluent Hebrew on my presentation.

After I returned from lunch, a young man introduced himself as Brian Goddard of UNESCO's Science Policy Division and asked me if I would be willing to act as a consultant for his organization. He and his boss were trying to develop a draft statement on the status of "professional workers," and they had run into all kinds of trouble with the representative of one of the Socialist republics, with their superiors, and with other UNESCO divisions. I enjoyed my sudden windfall of sociometric fortunes; as it came to pass, they were quite fleeting.

As the conference progressed, more and more time was taken up by discussions and less by papers. The first pointed exchange occurred between me and the tall, gray-haired, very cool Prof. Colin R. Austin from the Physiological Laboratory in England. Dr. Austin calmly challenged my notion that science should or could be guided, repeating the arguments I had often heard, advancing them in the stock way.

"I would like to make a few comments on the question of primacy of science which Dr. Etzioni mentioned," Austin said, opening fire. "I feel that in a way he has put the boot on the wrong foot here. It is not science that pushes society around, but the exploitations of science by industry, in the pursuit of profit; these are the real villains in the piece. And I think that is where the restriction should come, if it is to be made at all. In many connections I quite agree with him that restriction is entirely necessary. In other connections, such as, for instance, the question of sex control, it is vitally important that no suggestions be made that fundamental research should in any way be impeded. We need sex control, for instance, in the development of the agricultural industry, where it would be vitally important, and it should go ahead there, irrespective of its possible application to man. There are situations where its application to man might be dangerous, but at the same time it could solve a lot of human problems. And I think it is in that sphere of application where the curbing should be done, not in the initial determination of the basic possibility."

At first I wondered whether I should counter Austin or wait and see whether somebody else would pick up my side of the argument. I said to myself, "Don't allow the situation to be defined as a duel; let others get involved." But as Austin went on, my temperature rose, and my efforts to restrain myself melted. Just before Austin concluded, I indicated to the moderator that I wanted the floor next, and she let me speak.

With some heat, I began: "This is precisely the issue; it was raised most dramatically by physicists working on nuclear weapons, but it applies fully here. Is the scientist not to share in the responsibility of how the new tools he creates are to be used?"

"As much as any citizen, in his citizen capacity," Austin interjected.

"No," I retorted. "More so. *No* tools are neutral; it's like leaving long knives among small children and saying, 'Well, let them decide what to do with them.' The society has an immature mind, and is unable to digest complicated information and make complex decisions. The scientist, both because he has more information and because he causes the problem, has an extra responsibility."

"Would you agree," asked Prof. G. R. Fraser of the Department of Human Genetics, University of Leiden, Netherlands, "that this is a matter of controlling technology—not science? Could not one curb the *applications* rather than basic research?"

"I wish I could agree," I replied; and since I'm never able to answer succinctly, I continued: "But unfortunately there are now many scientific findings that short-circuit technological steps. Thus, once the formula for LSD is published in a scientific journal, almost any student of chemistry can make it in his basement—*without* any additional technological development. Similarly, I am told by my colleagues in physics, it will be possible in the foreseeable future to make A-bombs in high-school labs on the basis of published data."

I went on. "Even more important—and again I won't have the time to document this, but if I'm challenged, I will—it can be empirically demonstrated that the scientific process is not random. It's not true that as we conduct research at any one point we could make an unexpected turn and come up with a finding we never prepared for. Ninety-nine point nine percent of the findings are in the area we investigate. So when scientists work in astronomy, they come up with very few findings that are of great medical significance. And people who work in physiology are usually not very helpful when it comes to studying solar eclipses. Practically all findings are in the area of specialization. Now there are a few exceptions we can point to, but we should not try to make generalizations from the exceptions. That means that science is, in effect, already guided by society. Decisions are made in committees about where to invest resources, in what areas to train new Ph.D.'s. Thus committees determine where most of the scientific energy will be directed. So science direction is not random to begin with. Therefore, to say in effect, 'Why don't you worry more about the physiology of nutrition and less about the structure of crystals?' or in some other way help the scientific community make moral choices, is not a new intervention. Rather it's another vector which will influence the choices made anyhow. In my country, we are now moving research funds from space and defense to domestic problems—cancer research and the like. While a cancer researcher may come up with a new bacteriological weapon, studies indicate that this is much less likely than is often suggested. Thus science can be prodded as to where it is to focus, and above all, where it better not tread.... Please forgive me for being so long-winded."

Scientific meetings, especially those fraught with emotionally laden issues, rarely proceed systematically to any summary conclusion. Austin was now repeating what was, for him, an article of faith: "The scientist must be free to follow any lead; he might always stumble unexpectedly on something new in a completely unrelated field."

Reluctant to take the floor again, but also unable to keep my peace, I mumbled to Prof. Robert Reichardt, who was seated next to me: "If this is so, let them all research socially important areas, and let their findings reach surreptitiously into other areas, such as subliminal communications...." He nodded noncommittally.

But I was not alone. Prof. T. M. Fliedner, a young German pediatrician of the Institute of Clinical Physiology, University of Ulm, now spoke up. He began by explaining that, as a physiologist, he was interested in the feedback and regulatory systems of the body and their parallel in the society. (He was already winning my affection, because few things delight me more in such a meeting than finding another mind approaching the core issues from the perspective I hold valid.)

He then added: "I think the question Dr. Etzioni has raised has not really been answered: What are the systems, what are the institutions, that could effectively observe these developments and put out a warning finger? I think that the simple fact that CIOMS has been created by UNESCO and WHO, has probably been a reaction to the fact that people were concerned that some of these things may get out of hand. But I think it is equally important that we develop a sense of responsibility, of sensors, for the improvement of the recognition of ethical and moral values. We transmit from generation to generation an increasing amount of knowledge and technological know-how, but where is the evolution of the sense of responsibility in dealing with these advancements of science? I think a couple of hundred years ago, if somebody went mad and got a knife and killed somebody else, the implication for the total society was not so important. But if somebody today gets out of hand, he may drop an atomic bomb, and the implications for society are different."

I was worried that Fliedner, like Gellhorn, would end up endorsing a sentimental, but not viable, "sense of responsibility" which, God knows, we need, but also one which requires a set of institutions to back it up. Therefore I was especially attentive to his next words:

"So my question," Fliedner said in closing, "is when are we going to discuss the regulatory mechanisms? I think in this body we have to devise and develop methods of observing developments and keeping them in balance." Brief but very much to the point.

The moderator observed: "Now this is one of the main problems, the know-how of the know-how. I think our chairman wants to speak now."

Gellhorn took the floor: "I would like to speak—not as a chairman but as a participant in the discussion and mine is really a ramification of Professor Fliedner's position. I believe that Etzioni has been in the provocative sense attempting to emphasize characteristics of the two cultures—science on one hand and the good people on the other hand—and that the scientists have to become more aware of their social responsibilities. I believe that Etzioni would be the first one—because of his wide association with people in science, whether they be in physical science or the biomedical sciences—to recognize that the biomedical scientists or those in any other branch of science are people, and there are some who share his ethical position and some who do not. I personally would deny that the scientists have any special role in molding the application of their work to social ends. They have the same responsibilities as anyone else."

Gellhorn's comments evoked in me several feelings. I agreed with his statement that scientists had to become more aware of their responsibilities. But unfortunately, Gellhorn was in error on another point: scientists *of any kind* have additional responsibilities because they are more educated, more informed, and—to repeat—they created these problems in the first place.

Gellhorn continued "I would merely like to indicate one model, one perhaps unimportant model, that we are attempting in at least one school of medicine. Recognizing that health is more than physical well-being, we are attempting in our educational process —and it is not only for the students but for the faculty as well—to incorporate social scientists as part of our medical faculty. They are an intimate part, and they participate in essentially all the activities. From the moment that a student enters the anatomical laboratories to begin his dissection, there are social scientists there to indicate the implication of this unusual gift that society has given to medical students to study the body of another human being. It is not just stress on the proper identification of the nerves, muscles, blood vessels, and so on that is brought to bear, but also emphasis on the responsibility that is thrust on those who are going into medicine."

Did Gellhorn think I was plugging for social scientists, or that their incorporation into the medical school would make the students more responsible?

Gellhorn went on to explain that he and his colleagues were introducing social scientists on the clinical level. Medical and social scientists were studying the problem of obesity together, he reported. Gellhorn saw this "as a model that may work toward the answer that Professor Fliedner is seeking." He added: "And, further, I suppose that it is permissible to point out that the problem is not only in the biomedical area; for, as I have observed in France and as we have certainly observed in America, isn't it too bad that those who make automobile roads don't have a greater sense of their responsibility and what they are doing to our landscape? Perhaps if we had environmentalists working with road builders, this type of conjunction would begin to modulate and control some of the excesses which we are observing in modern life."

Austin was not going to let this stand: "I think the point really is that we should distinguish clearly between two areas of responsibility," he shot back. "There is the responsibility of the basic scientist to pursue knowledge for its own sake, without any regard to any other consideration, application, or what have you. There is the second responsibility, also perhaps of the same man, and from what we're saying today we would agree to be the same man, to consider the possible application. Well, these are two quite separate areas of responsibility, and I think that it is most important that basic research should be permitted to continue without any restrictions at all, in its own sphere and for its own sake, because at no time can you say that the application is going to be either good or bad. Knowledge in itself is neither good nor bad. It is the application. If the immediate purpose of the research is, as it were, a bad application, then this is unfortunate; but in the long run some effects of good may well outbalance the bad. We cannot look into the future. Therefore, we must keep these two areas of responsibilities quite separate, and I would reaffirm that the basic scientist should not be restrained in any way at all when he is concerned with his essential research."

No one could complain that Austin was concealing his position or that the viewpoint, "scientific freedom as an absolute, or superior, value," was underrepresented in the meeting. I raised my hand to indicate I wanted the floor, but put it down when I saw Fliedner asking for it. He did not disappoint me.

"Although I agree in principle with this statement," Fliedner said, "I must say that we have then to discuss who makes the decision as to how much this basic scientist gets in order to pursue his research, if you say he should not have any restraints. One of the strongest restraints I can think of is money. And who makes the decision on how much money is going to be spent on basic sciences in relation to other fields of applied sciences? I find this an extremely difficult question, and I have no answer myself; but I think this comes down again to what we discussed this morning: what are the institutions, the regulating mechanisms?"

I observed, briefly this time, that as the scientists participate in the allocative decisions, they can affect the allocation of the funds to less or more socially responsible uses.

But Austin was far from finished. "I wonder if it might help to distinguish between basic science and goal-oriented science or research," he said. "In the latter, the governing group that provides the funds obviously shapes the course of the work and controls what particular topics are investigated. But in true basic research, the scientist follows his nose. It is curiosity that leads him on, and he often does this with very little in the way of funds."

A light, doubting chuckle crossed the hall—at least I thought I heard it. Austin continued:

"And I'm arguing the case for the basic researcher to be quite free. He should not consider any possible implications; otherwise, it will influence his research. His first responsibility is to his science, to make his observations accurately and to record them, and to pursue this for the sake of pure knowledge. Later on, certainly, when it comes to an application, he may well come into the picture again and make a moral judgment on the work. But it is essential that the basic research be free."

Gellhorn now expressed what I thought many must feel: "I would accept that there may be instances where the basic scientist needs very little support, but I think that this is progressively less and less so." He added: "With regard to Professor Etzioni's suggestion that it is the responsibility of the scientists to aid in the formulation of the allocation of resources, I mention the fact that contemporary history indicates that it is society that determines where the research money allocation is to go. Thus, in the area we are discussing, the biomedical area, it has been the public that has said 'We wish to have cancer cured,' and that started more than thirty years ago. It started at a time when those persons doing cancer research recognized how little they knew, and the possibility of achieving a solution to the problem of cancer was very remote indeed. Regardless of this, cancer research has been one of the major areas that has received support. It is only, I would submit, within the past few years that there has been real enthusiasm on the part of the scientists in the biomedical community for such problems as cancer, where now there is at least a suggestion that we may be able to make fundamental contributions to increase our understanding and really get at the problem of cancer. But, prior to all that, regardless of what the scientists said, the public said that 'we wish to have cancer studied.' So I suppose my thesis, perhaps in contradistinction to Professor Etzioni's, is that scientists are only people; they have some voice, but I don't believe that it is a controlling voice."

As the meeting broke up for coffee, I felt quite satisfied. While Gellhorn, again, had painted me into a more extreme position than I had actually taken, and on the issue of responsibility, was still pointing the finger at "society" and away from science, he did recognize that scientists should and could participate in making decisions on the allocation of public funds for the various research goals. That the "people" demanded more attention be paid to cancer cures than, say, to the composition of the moon, did not trouble me. While a situation could arise in which the people would put undue pressure on science to serve their immediate needs rather than first advance science itself, this did not seem to me to be our predicament; on the contrary, greater relevance and social responsibility were now of the essence.

A Man of God

When the meeting reconvened after the coffee break, several international luminaries, wearing dark suits and carrying white, steaming styrofoam cups and pieces of pastry, materialized to hear a paper by Henry Miller, M.D., Vice-Chancellor, University of Newcastle upon Tyne, England.

Miller was not present, so his paper was read for him. I figured I could read it later and skipped out to call a colleague in Paris, to see if we could get together. I returned to the hall just in time to hear the closing lines being read.

Next, the program indicated, was Prof. Jürgen Moltmann, from Evangelisch-Theologisches Seminar der Universität Tübingen, in West Germany. I was rather curious to see what position he was going to take. I had found, over years of activity in public affairs, that many men of the cloth were rather insensitive to issues of public affairs; on the other hand a few were better informed on them than anyone else I knew.

Moltmann began modestly: "As a theologian, I think it is fair to say, first of all, that I don't have at hand divine answers for the new ethical questions in medicine." Next he spoke about what he called "the new ethical situation." He said: "The greater the medical power over vital processes, the more responsible are all those involved. In former times, health and sickness, life and death, were regulated by nature and destiny. Men were resigned to the fact that there was only very little to be changed. They took nature as an executive of the ultimate will and submitted to their destiny. When the Hippocratic oath ordered physicians to preserve life, it was life within the limits of opaque nature and destiny. More and more, the amazing progress of human medicine is today repressing the dependence on nature and is enlarging the realm of human possibilities to correct and change human conditions. Man becomes lord of nature and a steersman of his own vital processes. We speak, as never before, of health policy, population policy, and shall also combine 'genetic' with the word 'policy.' "

So far, so good; Moltmann recognized the active nature of man, unlike some theologians who see us as subject to nature, which reveals God's will. Yet I was waiting. What implications would be drawn from his general metaphysical position? [20]

"If the regulation systems of nature are more and more replaced by medical and social systems," he went on, "we need ethical values in order to regulate our human systems in a truly human way. One cannot shift the responsibility for life and death to nature any longer. It is not the ethical question of whether or not to use the Pill—or other treatments—but *how* to use the Pill in a responsible way. The same holds true for abortion, because up to now nature has been the greatest abortionist—by miscarriages."

He had to be a Protestant, I thought, for his point was aimed directly at Catholics, who would oppose man making such decisions and intervening in God's work. Moltmann did not stop to allow me to finish my thought. "Responsibility is indivisible and reaches so far as man's power reaches; more and more, man has to do what nature did earlier. This new situation causes many idle hopes regarding medical possibilities, and at the same time, great anxiety about manipulation. I think nobody can unburden the physician from the decision in individual cases concerning risks, sacrifices, life, and death. But patients, and the whole of society, have to share his responsibility. If today we see that the tradition of public ethics is antiquated in the light of the new medical progress, there must be public discussions about the new questions and challenges arising from that medical progress, or there will be no progress of mankind at all."

Right on! That's the way to do it! Values must reassert their primacy over technical decisions, and medical decisions are technical. But would he leave the new responsibility only in the realm of public education, or also invest it with a new institutional force?

Moltmann did not disappoint me. He added: "Perhaps we need a kind of medical parliament on the national, or better still, on an international level, to make the decisions in a fair and honest way. As with legal and political decisions, we must gain a social consensus about ethical criteria for the new decisions which *have* to be made: What is human? What is inhuman?"

Now my heart was cheering him; this one was on the side of the angels! A parliament on medical issues was indeed what we needed, and societal consensus—not just of the practitioners—was exactly what must be evolved!

Such a parliament should deal specifically with the social and ethical consequences of new developments in genetics and medicine. This Health-Ethics Commission would facilitate the anticipation of social and moral issues raised by the new genetic and medical techniques, would encourage public awareness and debate, and finally, would provide a convenient tool with which to regulate often overenthusiastic scientists and medical professionals.

Moltmann turned to demonstrate his thesis: "We are unable to furnish dialysis and transplantations to all sufferers from chronic kidney diseases. The resources are limited and the physician has to decide who may live and who has to die. Though, of course, one has to press society for increasing resources, one cannot hope for a society of perfect medical care ... but where are the criteria for the selection of survivals? The physician, who has to decide in such cases, is confronted with a moral dilemma. Following outside criteria, he can try to assess the value of his patients, allowing those to live who have the greater chances to survive. Normally these are the stronger or younger patients. Or he can save those who are of greater social value. But does a positive judgment not imply a negative judgment for those who are condemned to die? Is it possible to evaluate human beings like things and goods? Don't we destroy our own integrity by such judgments?"

Moltmann's position, if somehow it could be made the order of the age, seemed to me to meet three essential considerations: to begin, it favored medical and genetic progress; secondly, it faced, rather than avoided, the dilemma posed by human "imperfections," and allowed the doctor to act without abandoning the quest for a world in which he would have to condemn fewer and fewer persons while he saved others. Finally, Moltmann sought to reintegrate man by bringing the medical techniques, and the decisions which they force, into the context of our total societal value system and choice-making. Alienation, the mark of modernity, is the fragmentation of persons, the separation of their goals from their means. As Marx's answer is to reunite the two by public ownership of the means of production, so Moltmann sought to forge a union of medical technology and decision-making in a new ethics and a broad sharing of responsibility.

Moltmann closed with a review of the decision to die. His position was not particularly novel; much has been said and written about this in recent years.[21] But his approach gave a coherent normative underpinning to the new position of death as a choice rather than as a natural event.

"My next point concerns the ethical criteria of life," Moltmann said. "Our ethical criteria derive from our understanding of life and human life. In former days, the vitality of life and the humanity of life coincided. To live meant to survive. Today, people can be kept alive without their being conscious of anything at all. We *do* need a new definition of life, shifting from the quantity of life to the quality of life. Longer life is not equal to more happiness, and it can lead to *empty* life. Now it is said that the human character of life lies in self-consciousness. In any case human life is experienced and accepted, and where life is no longer experienced or accepted, we speak of 'dead life.' Hence, it is not the duty of the physician just to preserve biological survival of the organism, but to serve the *humanity* of life. The health of the human person is not identical with the ability to function or with the capacity for work and enjoyment, which is participation in social production and consumption. It is more the human health of the person, shown in the ability to mourn and to suffer, to sacrifice for others, and to die with dignity. . . ."

Again, Moltmann did not leave the matter on the level of abstract generalizations but brought it down to specifics, to the requisite earthy decisions. He was also unafraid to expose himself by stating his position quite explicitly.

He said: "It follows, for the determination of the end of human life, that under given conditions the irreversible death of the brain must be regarded as the actual symbol of the end of human life. It is of growing importance for the human and cultural integration of new medical treatments that man become able to accept his life consciously, to suffer humanly, and to leave life with dignity, if it comes to death. *The order of the body must be integrated into an order of the whole person.* This is important for the single patient—integrating treatments psychically; and it is important for society as a whole—integrating medical progress ethically, and developing, perhaps, a medical culture of life. Thank you."

The first comment came from Klein: "I was very interested in

Professor Moltmann's exposition," he said, "but was, I must confess, also a little disappointed. I am very often asked, as a biologist, to give my opinion on moral questions, and so I would have just liked to hear a little bit about his own personal convictions; instead, we just got a sermon. For instance, he says nature is the greatest abortionist, which is quite true. However, there are two diametrically opposed points of view which you can adopt. Should you now oppose nature and apply the traditional moral attitude? Or should you now assist nature because you feel that you are destined to do so and have more capacities to make a final decision?"

Klein continued to explain that according to one view, we should decide which fetus will be aborted and which allowed to live, on the assumption that we are better able than nature to distinguish who should live and who should die. According to the opposite view, we should not intervene one way or the other, which would mean, among other things, allowing all mongoloid fetuses to be born.

I wondered if Klein, in typical French rhetorical style—he was speaking as if he were on a platform in the Place de la Concorde—did not unnecessarily sharpen the issue, raising a false dichotomy. Could we not leave to nature some things (e.g., sex choice) and act on others (e.g., give help to parents who choose to abort mongoloid fetuses)?

Klein continued. "These are all difficult questions, and I would have very much liked to hear about your personal convictions. Instead we get a multiple choice, like new students who have to check one of three items—are you 'for,' 'against,' or 'intermediate?' —and we don't know what *you* are really thinking about this problem."

Professor Moltmann responded calmly: "My point was simply that man has to take over the role that nature or destiny or fortune played in former times. His responsibility is growing, and he cannot just shift it to nature; he cannot work for the preservation of life and at the same time say that the question of death is a problem of nature. *He has to take up the burden of his responsibility.* I say this over and against a certain religious opinion that one should not enter into the realm of nature because nature manifests the will of God, or something like that—which was a discussion inside the Catholic Church over the Pill, for example. It is not a wonderful thing to have to bear all these responsibilities, but one has to assume them. This was my point. And to some extent, one has to improve on nature, or at least replace natural systems of population growth and sickness by medical and social systems. Are these systems better, or not as good, as a natural system? That depends on how we use our responsibility. It is not a question of whether to take the responsibility or not, but how to use it. That was my point."

There was a brief pause, as if the chairman and everyone else in the hall were silently asking: "Professor Klein, is this quite clear enough? Anything else you wish to ask?" But Klein, head slightly declined, was reordering his notes.

The next question returned the dialogue to Moltmann's point about a parliament as a focus for evolving the new societal consensus on the values to govern our decisions on life and death and health.

A Collective Conscience?

"I would like very much to tell the members of this meeting a story that impressed me very much," Prof. Jean Hamburger said. "At the International Society of Transplants, it was reported that about three years ago, in certain countries which I will not name, small notices began to appear in the newspaper offering great sums of money in exchange for human organs for transplants. This alarmed us very much, and we started looking for means by which we could prevent such things from happening. Then we faced the concrete difficulties that are common with international organizations such as the World Health Organization, whose members, as its representative has already said, are slaves of a number of rules which forbid them from becoming a ruling or advisory body. It is therefore impossible, in spite of the great willingness of the health organization and its general director, to condemn this kind of human-organs market.

"This is the reason why, with this example in mind, I would like to return to a suggestion made by Professor Etzioni this morning which has been endorsed by persons such as Professor Moltmann —about the possible future role of new organizations, perhaps on an international level, which may acquire a recognized moral prestige, and hopefully, also the capability to act."

I was all ears; the words "to act" did not go over in the hall as well as the preceding lines. "To act" smacks of legislation, government intervention, and both scientists and doctors probably fear nothing more than the heavy, insensitive hand of the government pointing out to them what to do, and quickly moving from intervening in matters in which regulation may well be needed (e.g., that of flesh banks) to areas in which it would be devastating (e.g., regarding what a scientist is allowed to, or must, find). In short, most scientists and doctors feel that the dangers of intervention are so much greater than its benefits that they scream harsh words whenever it is so much as implied. Moreover, both researchers and practitioners see themselves as open to moral persuasion, with no need for guiding mechanisms. They tend to view themselves as motivated by concern for the patient; regulation should be limited to those governed by self-interest.

I wondered if Hamburger would be affected by this subtle cue of displeasure. I had looked up his affiliation in the conference directory: "Professor Jean Hamburger, Directeur, Unité de Recherches Nephrologiques, Groupe Hospitalier Necker-Enfants Malades." The affiliation clarified nothing, so I watched Hamburger closely. He was a vigorous man, who held his pen as if it were a stick; he moved his hands with force and spoke emphatically, unhesitatingly. He seemed to be either unaware of cues or, more likely, was a man with a mind of his own. He was saying: "In view of the many meetings which occupy themselves with *morality* in medicine and biology, I ask myself in what measure we could now envision the possibility of creating an international permanent organization that reflects the concern about these numerous problems which we mention year after year in these kinds of meetings—an organization, to repeat, which rapidly acquires an audience and a moral prestige sufficient to make it serve as a guide to nations in matters concerning medicine and biology vis-à-vis the rapid progress of the kind of questions we have been referring to in this meeting."

When the session ended, I walked over to congratulate Hamburger on his point. I had come to the meeting rather committed to the notion that the moral and social issues involved in the future of scientific development, particularly in the new areas of biological and medical research, could not be left to only the individual practitioner or scientist. As Prof. Bernard Barber has written: "Clemenceau once remarked that war was much too important to be left to the military. In the same fashion, science and its consequences are much too important to be left to the scientists. In both cases, the instruments are much too important to our social purposes to be left wholly to the experts in using those instruments. They are the concern of all who have the responsibility for our social purposes." [22] I had previously called for the establishment of a commission of scientists, practitioners, humanists, and theologians to explore these matters, to lead public awareness and education in the issues involved, and to act as a collective voice of the social and moral values at stake.

On the way back to the hotel, I entrusted myself to the Métro, and as I rode, reviewed my previous efforts on behalf of a Health-Ethics Commission.

My first attempt to advance this position publicly was in 1968, in *Science*, the official publication of the American Association for the Advancement of Science. I chose *Science* as my platform because it has hundreds of thousands of scientist-subscribers from all over the world and because the popular press often picks up reports that *Science* carries. My choice turned out to be both effective and mistaken. The article I wrote did elicit more response than anything I had written before in a long career of writing: stories about my *Science* article were carried by the main American and European newspapers and weeklies, and even reached the *Vietnamese Times* and the *Sydney Morning Herald*.

However, they all focused on the issue I used for illustrative purposes and neglected the one I was seeking to illustrate. I discussed the evolving procedures for allowing parents to choose the sex of their yet-to-be-born child and highlighted the undesirable consequences of such a breakthrough. I thought that by using a "low-key" problem—rather than a more dramatic or complex one, such as thought-control or test-tube breeding of optimal specimens of babies—one could focus on the regulative issue which affected everyone: Is it necessary to regulate science, or can we continue to allow it to follow its own leads? Was the freedom of science an absolute value that took priority over all others? If not, under what conditions should others take precedence in determining our course? And how could the regulation of science be achieved, without damaging its vitality and its freedom of inquiry? I had also hoped that by focusing on a concrete and relatively delineated case instead of on, say, all genetic engineering, one could more readily deal with the general issues.

However, quite naturally, the press, letters to the editors, and the mail to my office and home all seized on the concrete issue—when and how could we order a boy? a girl? rather than on the regulative issue.

The effort, though, was far from wasted. Aside from netting me a measure of recognition in the community of those dealing with these issues (the *Science* article is still often cited, which is the way scientists pay each other recognition), the mail affected my position. While I never opposed sex choice (and other advances in genetic engineering), I did object to its development before its implications were thoroughly reviewed, that is, before the personal and societal costs and benefits were examined and the necessary precautions taken (e.g., one might favor the development of the technique but alert the public not to use it lightly). I was expecting that reviewers would recommend that we leave well enough alone. Nature's ratio of roughly fifty-one boys to forty-nine girls at birth is quite acceptable (greater male mortality during infancy, in wars, and from heart attacks results in later years, in a disproportionate number of women). If sex were to be determined by our choice, it would only increase our personal and collective burdens. My mail, though, raised some new questions. Obviously, the matter was of terrible concern to some people. A woman writing from Texas pleaded with me to tell her if I knew a way to ensure a boy; she said she had four girls and her husband had threatened to leave her if she had another one. (I am not sure that it helped much to explain to her that it is *his*, not her, contribution that determines the sex of the unborn child.)

Others have written in a similar, albeit somewhat less agitated, manner, seeking either a girl or a boy, because they were blessed with only one kind. For example, a gynecologist reports of a patient's husband: "He was head of an old family firm, proud of his lineage, and desperately wanted a boy to carry on his name and business. Three times straight, much to his dismay, I had to warn him *not* to expect a son, and each time I was right. The father shook his head after the last one, saying grimly: 'We are going to keep trying. I've got to have that boy.' " [23] The Shah of Iran divorced his first wife when male offspring were not forthcoming.

It seemed I had underestimated the unhappiness that sex choice could alleviate. Still, a systematic review was needed. Would the aggregate of undesirable consequences for the public outweigh whatever personal gratification sex choice could offer? And would it generate more personal and familial conflicts (between parents who disagreed about what the next offspring ought to be) than solace for those who could have their choice? A review board may well first have to ask an even more elementary question—how does one assess the relative size, strength, and merits of these various camps?

My next effort to focus public attention on the need for societal guidance of scientific developments was not much more successful than the first one, but for a quite different reason. It was undertaken when Rabbi Marc H. Tanenbaum, of the American Jewish Committee, asked me to address a meeting of rabbis and ministers conducted jointly by his committee and an organization called The Council on Theological Education and the Commission on Ecumenical Mission and Relations of the United Presbyterian Church. On February 9, 1970, therefore, I found myself in Princeton, New Jersey, in the assembly room of one of those roadside motels, talking up my old idea. I did not keep my notes from that evening, but according to the *Trenton Evening Times*, I said that "society must regulate science to prevent scientific progress from overwhelming man," and "called for a national council of religious leaders to work together with the scientific community to protect man and his values from an escalating flow of scientific innovations" that may soon include, "new drugs which modify behavior, subliminal advertising, opening of the genetic code to human manipulation, and research purporting to establish racial inferiority." I also said, "Religious leaders should join the dialogue and guidance efforts, not as self-appointed czars, but as spokesmen for the moral concerns which a science, unleashed and gone wild, tends to overrun."

The *St. Louis Globe-Democrat* quoted me as adding that "our society has proceeded on the assumption that science should be free to investigate any lead it wishes to follow, and that any new result will be allowed to spill over freely into society, and society will have to adapt. But we can no longer accept scientific claims to an unlimited, unqualified superiority over all other values."

Well, I have given many talks in my lifetime to a large variety of groups; so I have learned to read the audience. Some audiences get turned on; others are outright hostile. My Princeton audience was, to my surprise, indifferent, remote, or dubious. The questions which followed my talk were few and brief. Kenneth Vaux,

Professor of Ethics at the Texas Medical Center's Institute of Religion would go only as far as to agree that manipulation of sex and other physical traits posed "serious ethical questions." Msgr. Marvin Bordelon, Director of the Department of International Affairs of the United States Catholic Conference, said it is "good and important" that these problems are tackled, but he was skeptical about a council. "More meetings would not solve anything."

Above all, the participants much preferred the higher reaches of metaphysics and supergeneralities to any discussion of a role *they* might play. Dr. Hans-Ruedi Weber, Associate Director of the Ecumenical Institute of the World Council of Churches, felt that "the acceleration of scientific discoveries has triggered movements which have brought into profound spiritual crises all traditional ethics, spiritualities, and faith," and went on from there to talk about the economy of abundance and East-West tensions. Prof. Lionel Rubinoff of York University asked: "How is it possible to teach the sanctity of life and the sacredness of the individual in a world that is becoming more and more subject to control, balance, and order?"

I took the last bus home from Princeton, got to bed rather late, and woke up tired the next morning, quite sure I had gotten nowhere. It was one of those rare occasions when I wondered why one bothers—a feeling which, for me, never lasts very long. So I was cheered when the next opportunity presented itself. I read in the *New York Times* that in a hearing before the House Appropriation Subcommittee on the National Institutes of Health, Dr. Joshua Lederberg, a Nobel laureate, had urged the establishment of a National Genetics Task Force to increase the momentum of efforts aimed at unlocking the genetic code of man. Such a breakthrough, Lederberg had explained, could lead to the prevention of many illnesses whose origin is wholly or partially in the genetic code.

I responded by expressing my concern on the editorial page of the *Times*. In my article, which appeared on September 5, 1970, I wrote:

> There is much to be said in favor of such a task force. But it ought to be accompanied by a task force on the social and moral consequences of genetic manipulation. The imminent breakthroughs in biology may affect man as much or more as he was affected by previous revolutions in engineering and physics: the imposition of a new set of capacities, of freedoms, of choices society must make, of the evil it can inflict.

Gene manipulation may also allow man to tamper with biological elements which heretofore had to be accepted, including the sex of children to be conceived, their features and color, and ultimately their race, energy levels, and perhaps even their IQ's. Thus, what may start as the biological control of illnesses could become an attempt to breed supermen. While this may appeal to some, think about the agonizing problems if man has to act as the creator and fashion the image of man.

Fortunately, it seems we do not have to stop the genetic combat of illness to prevent genetic engineering for racist purposes. . . . One kind of genetic manipulation will not willingly open the door to others.

Actually, most scientific findings are not readily transferable, and their application is affected by moral taboos.

Before such guiding of scientific efforts can be effectively applied to the new genetics, we must have a clearer notion of the moral and social choices involved in the biological revolution and the mechanisms by which science can be guided without being stifled.

Let us not again sail blindly into a storm unleashed by scientists anxious to unlock all of nature's secrets, with little concern for who and what will be blown over in the resulting tidal waves.

To this end, I suggest that at least 1 percent of the $10-million a year requested for a National Genetics Task Force be set aside to explore the options genetic engineering is about to impose on us.

Lederberg, whom I knew from a brief encounter in Washington, D.C., is not a man plagued by doubts or one to mince words. He came back at me, in a letter to the editor of the *New York Times:*

Professor Amitai Etzioni's September 5 Topics column "Genetic Manipulation and Morality" is another contribution to the demonology of genetic engineering that obscures the important dilemmas of health policy requiring open-ended public discussion and participation.

The Congressional committee testimony to which he alludes gives no justification for "ordering superman" by the task force which I advocated. It is a plea for establishing the relative urgency of various categories of human misery, like mental retardation, cystic fibrosis, heart disease, diabetes, and many other conditions which have an important genetic component.

"Shopping for genes" [a subtitle the editor inserted in the middle of my article] is a phrase of his own invention; perhaps he means nothing more than the aspiration to a healthy life to which most of us plead guilty. I also stressed that it would be both technically and socially advantageous to concentrate on ways of modulating the untoward effects of deleterious genes whenever possible for therapeutic purposes in preference to strenuous efforts at modifying the genes themselves.

His point that we ought to explore the aggregate social effects of individual decisions is an excellent one. This of course is important in the therapy of genetic disease, but also in the assessment of every other claim on precious resources like scientific talent and medical service. It is especially needed when humanitarian motives lead us out of the conventional marketplace, where each consumer makes his own allocation of limited resources for the most valued aims.

Such an exploration is, however, confused rather than advanced by phrases like "genetic engineering," which are as prejudicial as it would be to call surgery "anatomical manipulation," education "psychological control," or scientific nutrition "molding a superbaby."[24]

The letter carried the note that the author, Professor of Genetics at Stanford School of Medicine, had won the Nobel Prize in physiology and medicine in 1958.

While the letter's strong critical words, coming from one of the leading authorities in the field, could not but displease me, I was gratified to see that the dialogue was extended and to note—on second reading, I admit—that the difference between Lederberg's position and my own was much smaller than the letter's sound and fury initially suggested. Lederberg did see dangers in genetic interventions and recognized a need for review; that he was somewhat defensive about his discipline was to be expected.

I was much more troubled by an individual who wrote to the *Times*:

Being a father of two sons who are infirm with muscular dystrophy, I found the September 5 Topics column by A. Etzioni on genetic manipulation repulsive and totally devoid of sense. I see my two sons withering and weakening as the days go by. The future holds nothing for them but the inevitable wheelchair at nine and certain slow death by nineteen. The only scant hope for their survival is the effort made by medical research to break the genetic code and be able to reverse the disease.

Comes Mr. Etzioni and advocates a go-slow attitude toward accomplishing this scientific feat, until he and other professors of sociology make up their mind, "What supermen will the national task force order? Blond or brown, white or black?" To satisfy Mr. Etzioni's intellectual imagination, I shall answer him: Just revive dying children first, produce supermen later.

Mr. Etzioni's suggestion ". . . to set aside at least 1 percent of the $10 million to explore the options genetic engineering is about to impose on us" should be indignantly rejected. Our government is incredibly tight in providing the pennies which can be spared in the budget for medical research and for saving the lives of dying children.

It would be a sheer tragedy to waste them on idle sociology professors to explore options created in their imagination. [25]

At first I felt so overwhelmed by the personal agony of the sons and their father that I did feel nothing should be done to delay service to them. The easy way out was to suggest that the funds the ethical review required not be deducted but added to those available to genetic research. On a deeper level, I eventually realized that there is no way to rush a service to some without inflicting greater misery on more, as the rushed enactment of PKU legislation suggested.

Legislation requiring the testing and treatment of newborns for PKU was the product of a campaign of lay pressure supported by what some consider to be inconclusive and inadequate scientific research. Those pushing for the legislation were mainly parents of afflicted children and others intimately affected by the disease or by other birth defects. Their crusade was accelerated by the enthusiastic support of physicians and researchers of congenital metabolic disorders. These individuals hoped that compulsory testing for PKU would turn up more cases and would further their knowledge of other genetic and metabolic defects.

The emotional intensity of the proponents and their eagerness to establish some sort of defense against the disease, as well as the nature of the legislative procedure itself, hastened the legislation, blocked opposition, and discouraged any delay to establish more conclusive evidence that the test and the treatment were reliable. As of 1971, forty-three states had some sort of legal provision for the test, and implicitly or explicitly, for the treatment of PKU.[26] (Also see pages 24–25, 105.)

No rational decision can be made if the agony of an individual—however moving or tragic—is its basis, and the greater sufferings of the greater number—which are less immediate and less dramatic—are ignored. True, it is the essence of our libertarian tradition not to sacrifice a person for the multitude. But one cannot disregard, either, that the multitudes are made out of nothing but many individuals, all created in God's image.

Above all, the quest must be for solutions which allow those afflicted to be served while protecting the rest of us, rather than seeking either to block the progress of genetics or to embrace all its offspring indiscriminately.

This time I was not left without support. Peter Steinfels, a young Catholic author whose work I had previously followed in *Commonweal*, wrote in reference to my article and to reports that genetics will help people have "optimal babies":

The recent proposal for a National Genetics Task Force, for example, aims at preventing genetically based illnesses, and I have heard it argued that medicine based on a fundamental knowledge of the genetic code would make today's medicine look as primitive and foolish as the doctors of Molière. One wonders, however, whether the backers of such proposals are reading the memoirs of nuclear scientists who, on the 25th anniversary of the explosion of the Bomb, are reflecting on the awesome consequences of their simple desire to advance science and defeat Hitler.

Is there any way our society can avoid being led, step by step, into a situation for which it is entirely unprepared? Etzioni thinks so. He is optimistic about the possibility of channeling genetic engineering into certain fields, such as the prevention of illness, without its being harnessed for other ends, like racial domination. He argues that most scientific findings are not as transferable as people think, and that "their application is affected by moral taboos." For example, "those scientists who sought to prove racist theories are starved for funds and academic recognition." Without sharing his optimism, I certainly approve of his warning that we cannot guide these scientific efforts without a better idea of the moral and social choices they imply, and that a proportion of any funds spent on genetic engineering be devoted to that purpose.

. . . In the long run, we must face the issue of the stoppability of science and technology, and I would propose a National Stoppability Task Force to investigate whether science is stoppable at all, under any conditions, and if it is, then exactly how is the miracle accomplished. Has there been a *single* example of scientific research abandoned—when it was producing good results—because of moral reservations about its consequences? That feeling of despair will hardly be relieved if the Senate approves the appropriation for the SST. "Rarely will so many be bothered . . . to save so little time for so few," writes Henry C. Wallich of the supersonic jet in *Newsweek*. Yet Wallich insists that "we cannot escape" building the plane—for economic reasons. Rarely, one might add, have so many reports and studies indicated so many reasons against doing something to so little avail. Stoppability should begin here. [27]

In 1971 Sen. Walter Mondale, one of the most liberal members of the U.S. Senate and the one closest to social science circles, conducted hearings on a resolution to establish a National Advisory Commission on Health, Science, and Society. The new body, to act as a congressional rather than presidential commission, was to study and focus national attention on ethical and policy questions raised by new advances in biology and medicine. The body was to serve for two years, command a budget "not exceeding one million dollars a year," and be directed by a board of fifteen leaders in the fields of law, theology, medicine, government, and the humanities. The commission was to be backed up by research, to be conducted by its staff and by outside sources (for the text of the bill, see Appendix 6).

The various scientists who testified or sent testimonies with reference to the bill ranged from those strongly in favor of such a resolution (as was Dr. James D. Watson, discoverer of the DNA) to those who saw no need for it (e.g., geneticist Arthur Kornberg). I put in my two bits. I realized, of course, that a two-year "study commission" was not sufficient; the nation needs a permanent commission and one which, besides "studying problems," would also be charged with formulating alternative guidelines for public policy. Also, I saw a need for the national commission to be backed up by a myriad of local review boards. The national commission would deal with policy guidelines (e.g., should the use of amniocentesis be encouraged or discouraged?), and would even run experiments—a kind of FDA for medical procedures, only without coercive power. The local boards would review individual decisions (e.g., was a particular physician correct in refusing to provide artificial insemination to a couple he deemed "neurotic"?), and attack specific problems (e.g., making arrangements to set up genetic counseling in parts of the country where specialists are not available). Mondale's bill did not provide for all these issues, but it was, I felt, a fine start.

The Nixon administration's position was presented by Dr. Merlin K. DuVal, Assistant Secretary of Health, Education, and Welfare, who maintained that other groups were already studying the matter. Reference was made to both a study conducted by the Committee on Life Sciences and Social Policy of the National Academy of Sciences, under the direction of Dr. Leon Kass, and to studies of the

Institute of Society, Ethics, and Life Sciences of Hastings-on-Hudson, headed by a young Catholic lay theologian named Daniel Callahan. I felt that though these bodies were important, and their studies very thorough and useful, [28] they lacked the authority and the limelight; hence a national commission set up by Congress would have a much stronger effect.

I was elated when the Senate in December 1971 unanimously endorsed Mondale's resolution, which was at the time co-sponsored by about twenty senators from both parties. Several press reports which followed talked optimistically about the commission as if it were about to start its deliberations. But having seen Congress at work before, I realized that a House act had to follow, to be followed in turn by a joint conference if there was a difference in the two versions of the bill, a presidential signature, authorization of funds, and a few other steps. The House did not act, when the Senate did, in the first months of 1972. Actually, just before I left for Paris at the end of August 1972, the bill had not yet been passed by the House. So I had called Mondale's administrative assistant, Mr. Herman Jasper, who explained to me that the bill was in a House Subcommittee on Health and Environment, chaired by Congressman Paul Rogers, Jr. (D.-Florida). "They are not holding it up but just did not get to it. Though if it is not acted upon soon, and Congress completes its session, the whole process will have to start from the beginning." Concerned, I called Rogers' administrative assistant, Mr. Stephen Lawton, who also stressed that Congressman Rogers had nothing against the bill, "but we must hold hearings on it; can't just accept it; and there are many other bills to be acted upon."

I left for Paris quite pessimistic over the fate of a resolution very much needed but not backed up by either the Administration, a lobby, a civic group, or a great public outcry. But one does not give up. Now, a day into this international meeting, I saw a new sign of life. The fact that Moltmann from West Germany, Hamburger from France, and Gellhorn from the United States had all expressed an interest in such a Health-Ethics Commission offered a new chance. Regardless of what happened to the American commission, could there be an international foundation to express a collective voice on these matters for which the need for concern, reflection, and pronouncement of new values, (or redefinition of old ones) was all too evident?

Part II:
THE SECOND DAY

CHAPTER TWO

A Spokesman for Test-Tube Babies

The next morning, the drama heightened. Dr. Austin presented a paper about the "initiation of human development *in vitro* and transfer of early embryos." No other recent development in biological engineering has raised as much doubt among the public as that involving experiments in which conception has been carried out, and gestation fostered, in a test tube, which is what *in vitro* ("in glass") means. The press, somewhat ahead of the scientists, dubbed the development "test-tube babies." While no fetus has yet been carried to full term in a laboratory, many believe such a development is at hand.

The basic procedure entails the removal of an egg from a woman and the fertilization of it with sperm. One such resulting human embryo has lived seven to eight days *in a test tube.* Doctors have also attempted to transplant fertilized eggs back into the wombs of women with blocked Fallopian tubes who could not conceive in the "old" way. Although these replantations have not been successful, experts agree that it is just a matter of time before they are and before it will also be possible to plant laboratory-fertilized eggs in the wombs of host mothers or "mercenaries," as one biologist referred to them.

Work is also in progress on an artificial womb for growing such babies "from sperm to term" in the laboratory. Prospective mothers could then simply contribute an egg, and avoid pregnancy and labor. The next step—though quite remote—is the production of identical copies of a person, called "cloning," which entails asexual reproduction. This is achieved by stimulating one of the body's cells, and it has already been done with carrots and frogs. [1]

The moral, social, and legal issues raised by these present and prospective developments are many and complex, and I had never before had a chance or a reason to form a definitive position on them. As I'd read about these studies in various popular and professional publications, I had mentally filed them under "too complex for a snap judgment." And, indeed, I was happy to see that others were similarly reluctant to make quick decisions. Shortly before the Paris meeting, the *Journal of the American Medical Association* published an editorial asking for a moratorium on experimentation and study of *in vitro* babies.[2] The editorial stated that "the time seems clearly at hand" to stop experiments until physicians, scientists, philosophers, and theologians had a chance to review "the thorny issues raised by genetic engineers." The editorial was backed up by two articles by Dr. Paul Ramsey, who is Harrington Spear Paine Professor of Religion at Princeton University and a leading Protestant theologian.

Elsewhere, two Nobel laureates, Dr. James D. Watson (of DNA fame), and Dr. Max Perutz, a senior British scientist, attacked in strong terms the work carried out by Austin and his colleague, Dr. R. G. Edwards; at the very least, they said, these new experiments had to be carefully scrutinized. Watson called it "a matter far too important to be left solely in the hands of the scientific and medical communities."[3] Perutz called it a "stunt" and suggested that "the whole nation should decide whether or not these experiments should continue."[4]

Now, as Austin was being introduced by Gellhorn, I reexamined an article I had brought with me to Paris which very systematically explored the issues raised by the test-tube-babies experiments. The author, Leon R. Kass, an M.D. and a molecular biologist, gave the experiments and their numerous implications a thorough analysis which must have taken months of meticulous research and deliberation.[5] He wrote: "I had earlier raised the question of whether we have sufficient wisdom to embark upon new ways for making babies, on an individual scale as well as in the mass." And he concluded: "By now it should be clear that I believe the answer must be a resounding 'no.' To have developed to the point of introduction such massive powers, with so little deliberation over the desirability of their use, can hardly be regarded as evidence of wisdom."

I now waited anxiously to learn more about *in vitro* fertilization from its chief practitioner. Austin, Edwards, and their team are not only carrying out the headline-making studies; they also serve as spokesmen and advocates for *in vitro* research. Like Austin, Edwards stresses the value of their basic research rather than claiming the sole purpose is medical service to infertile women. He recently stated: "We are well aware that this work presents challenges to a number of established social and ethical concepts. In our opinion the emphasis should be on the rewards that the work promises in fundamental knowledge and medicine.' [6]

As Austin began to read his paper, the hall was particularly silent; the usual informal exchanges between scientists seated next to each other were noticeable by their absence.

"In more than one-third of the infertile marriages investigated, the cause has been found to be sterility in the woman due to actual occlusion of, or functional or anatomical anomaly in, the oviducts, and surgical measures and drug therapy are effective in less than twenty percent of the cases," Austin began. He went on to say that the treatment of infertile women by his colleagues Dr. R. G. Edwards and Dr. P. C. Steptoe "is undertaken in the conviction that it is a basic human need, even a 'right,' to have a family, and that the threat of overpopulation does not justify refusing aid to the infertile, any more than to the undernourished. Nor is adoption a satisfactory alternative, except to a small minority of infertile couples."

Here Austin was indirectly responding to the criticisms of Kass and others, who complain that less controversial alternatives had not first been exhausted in the treatment of infertile women. Kass asked: "Were they told about alternative possibilities, such as surgery on the blocked oviduct or adoption? Since ... three out of forty-six 'infertile' women [in these studies] became pregnant—by the 'old,' customary method—during the first month after the laparoscopy [surgical procedure used by Austin *et al.* to obtain eggs for *in vitro* fertilization] we can only wonder about the criteria used for subject selection." Austin's response to his critics seemed quite reasonable to me. It is easy to see that for at least some infertile women no other procedure would help. At the same time, Austin did not explicitly state that everything else was tried first for the specific women he used for his studies.

Instead, he moved on to explain the details of the preparation of the women for the test. "Although the patients concerned in this work are primarily handicapped by occluded or pathological oviducts, they almost all have regular cycles with complete follicle development and spontaneous ovulation. To facilitate the acquisition of fresh normal oocytes at a stage awaiting fertilization, the patients routinely receive hormone treatment."

An oocyte is an egg cell that has not yet fully matured. The hormonal treatment requires the use of two hormones; one is called "human menopausal gonadotrophin (HMG), which is mainly follicle-stimulating in its action"; the other is "human chorionic gonadotrophin (HCG), which is strongly luteinizing and accelerates the advent of ovulation." The luteinizing hormone (LH) is required for the development of the follicle, or the sac, which surrounds the egg.

It occurred to me that these careful words do not capture the full scope of the procedure used. Women are injected to cause them to "superovulate" (to produce many eggs) because, when left untreated, they produce only one mature egg per menstrual cycle. I wondered if there was a medical reason to induce in these women this unnatural and possibly harmful condition—or was the goal simply to get more eggs out of the "volunteers" in a short time? And what were the risks involved?

Austin continued to unfold a description of his work, discussing the way the eggs are obtained. "Access to the ovary in the patient is achieved by means of laparoscopy.[7] The laparoscope is essentially a thin telescopic instrument, equipped with 'cold' lighting by means of glass-fiber optics. It is passed through the abdominal wall in the region of the navel where large blood vessels are lacking, and permits close examination of parts of the abdominal walls, or viscera, on which it can be focused so as to give an enlarged and detailed image. Manipulating instruments of even narrower diameter can also be introduced through the abdominal wall, and aid inspection by permitting positioning of organs or make possible a variety of surgical operations.

"Laparoscopy can be done with a local anesthetic, but general anesthesia is preferred because of better abdominal relaxation. The patient is commonly placed in the Trendelenburg position (she lies supine on an inclined plane with the pelvis higher than the head), and the abdomen is inflated with carbon dioxide, nitrous oxide, oxygen or carbon dioxide in air to increase maneuvering space within. A small incision is made in the navel to admit the laparoscope, and is closed with a suture after withdrawal."

Many physicians would hold that any general anesthetic and any surgical procedure involve an element of risk. Hence the question of the legitimacy of these experiments can be raised on this ground alone. It seemed to me, though, that the question of risk was not unique to these particular experiments. Hence if *in vitro* was otherwise "OK," and such experiments were generally tolerated, the risk was not a reason for prohibition. At the same time, the fact that this line of experiments required surgical intervention did not make it one that could be easily favored.

Austin went on to spell out how test-tube mating took place once he had extracted the eggs. It's nothing like the natural process, to be sure: "All oocytes are placed in a carefully balanced fertilization medium into which the spermatozoa are introduced. After several hours incubation, the oocyte is examined *in situ* with a microscope. Evidence of fertilization (dissolution of cumulus-cell mass and presence of spermatozoa in the zona pellucida or in the perivitelline space) has been found in many oocytes examined several hours after insemination. The oocytes are then washed and transferred to a culture medium. Inspection of the developing embryo is made two or more times daily to determine whether cleavage is proceeding normally."

Critical reports on this work have dwelled on the fact that when the development of the test-tube fertilized egg proves abnormal, and hence when it becomes clear that the resulting child would be deformed, the fertilized eggs must be "washed down the sink."[8] One way these "laboratory abortions" have been justified is by pointing out that under normal conditions, early spontaneous abortions or miscarriages occur frequently and often for a similar reason—a genetic deformity. But what if the fetus is eight months old—and would of course be a premature, although quite fully formed baby—when the scientists discover that their product is defective and that the "parents" will refuse to accept delivery?

Austin himself is not in the business of growing test-tube babies for long periods; however, others are. For example, Dr. Daniele Petrucci of the University of Bologna "reported that after more than forty failures he had successfully fertilized a human egg *in vitro*, cultured the embryo for twenty-nine days ('a heartbeat was discernible') and then destroyed it because 'it became deformed and enlarged—a monstrosity.' "[9] On this, Ramsey has written: "Petrucci yielded to his Church's condemnation of producing a human being without 'the most supreme assistances of love, nature, and conscience' (editorial, *L'Osservatore Romano)* and became a forgettable episode in the history of *in vitro* fertilization research."[10] While Austin was not in Petrucci's shoes or lab, it soon became clear that he must face rather similar moral dilemmas.

Instead of trying to "grow," or cultivate, a fertilized egg in the laboratory, Austin says that his colleagues would like to replant the test-tube embryos in the uteruses of the egg donors. "The transfer of the 8-16 cell embryo into the uterus will probably be attempted by passage through the cervix. Experimental work with laboratory and farm animals has shown that the timing of introduction of embryos, relative to the normal time for the implantation of the embryo in the uterus, is critically important. This is because the hormonal balance in the female must be appropriate for embryo implantation. There are good prospects that when the human embryo is returned to the patient who provided the oocyte, her hormonal balance will in fact be appropriate; but this point has yet to be established. Extraneous hormone administration may yet prove necessary."

Austin turned next to examining the risks, several of which were apparent, when he depicted the basic procedure. He first spoke about the hormone treatment used to treat infertile women and to pepare patients for egg removals. "Women patients, presenting for reasons of functional infertility, have for many years been treated with HMG, followed or not by HCG, in the effort to prepare them for pregnancy initiation by intercourse or artificial insemination. The most serious risk would appear to arise from overstimulation of the ovary, causing massive follicle rupture and some hemorrhage; nowadays this is routinely avoided by initial testing of each patient's sensitivity to gonadotrophins."

Austin's reassurance was somewhat less than complete. Such tests are rarely, if ever, that perfect. [11]

The potential danger that Austin next discussed is the surgical procedure for removing the eggs from the ovaries, the laparoscopy. "The main risks in the application of laparoscopy are those attributable to the induction of general anesthesia, the establishment of the pneumoperitoneum, and the insertion of the laparoscope. Obviously much will depend upon the skill and experience of the anesthetist and the surgeon; in accomplished hands the chances of accidents and complications are in fact extremely low."

Austin went on to explain that the laparoscopic procedure does not necessarily damage the ovary or the many Graafian follicles, or sacs, which contain the egg cells (as well as important hormones). "Again, much depends on the skill of the surgeon and his assistants, but the injury normally inflicted—the puncture of one or more ripe follicles—would not greatly exceed that occurring with the natural rupture of follicles. Careful inspection of the ovary after oocyte collection minimizes the already small danger of significant hemorrhage."

Finally, regarding fertilization and culture *in vitro*, Austin stated: "The dangers here are those that could affect the life and well-being of the future child, and arise from the possibility that the manipulations to which the oocyte and embryo are subjected may provoke changes leading to *birth defect*. This could happen through disturbance of the terminal stages of meiosis, the normal oocyte reactions restricting the number of fertilizing spermatozoa, or the progress of the first or later mitotic divisions.* In experimental animals, intentional disturbance of these events has been shown to give rise to haploidy, triploidy, and tetraploidy in the developing embryo." (A haploid embryo would be one that has half the number of chromosomes normal for body cells. A triploid has three times the haploid number of chromosomes, or one-and-a-half times the normal number. A tetraploid has four haploid sets of chromosomes.)

Austin discussed these chromosomal disturbances: "The conditions are evidently highly lethal, for embryonic development terminates about halfway through pregnancy, and the birth of 'pure' haploid, triploid, or tetraploid animals has yet to be reliably recorded. A few spontaneously occurring and apparently 'pure' human triploids have been born but did not survive long. Several

---

*Mitotic division refers to the process of regular cell division. The result of mitosis is two identical nuclei having the same number and same kinds of chromosomes —Author.

instances of mosaicism* involving a proportion of heteroploid cells are on record in human subjects, some of whom were seemingly normal; mosaicism could also arise through disturbance of one or more mitotic divisions during the cleavage of the embryo. Except where mosaicism conceals a small proportion of abnormal cells, chromosomal defects can be readily diagnosed in the developing human fetus by the techniques of amniocentesis and the karyotyping [measuring and labeling a cell's chromosomes to see whether it deviates from a standard pattern] of cultured fetal cells. Pregnancies established in human subjects with embryos initiated *in vitro* would naturally be monitored by these and other methods."

Was Austin playing down the risks? Other scientists do not seem to pass over them so lightly. Kass, for example, pointed out: "The truth is that the risks are very much unknown. Although there have been no reports of gross deformities at birth following successful transfer in mice and rabbits, the number of animals so far produced in this way is much too small to exclude even a moderate risk of such deformities. In none of the research to date has the question of abnormalities been systematically investigated. No attempts have been made to detect defects which might appear at later times or lesser abnormalities apparent even at birth. In species more closely related to humans, e.g., in primates, successful *in vitro* fertilization has yet to be accomplished. The ability regularly to produce normal monkeys by this method would seem to be a minimum prerequisite for using the procedure in humans." [12]

I wondered why Austin and Edwards chose to work with humans when very little, if any, work had been done on other primates. [13] My colleague Dr. Barber suggested a possible reason: primates are expensive and difficult to get. Maybe when experimental animals become more easily available. . . .

Austin was now summing up. When all was said and done, the risks were not considerable, and he went on to point out, he provided for the possibility of a malfunction. "Questions of an ethical nature arise from the hormonal and surgical treatment of patients, and from the extracorporeal induction and culture of human embryos." Considering the small degree of detriment done to the patients, the preliminary hormone treatment, the laparoscopy, the removal of

---

*Mosaicism refers to a condition in which different cells or patches of tissue of unlike genetic constitution are mingled in the same individual—Author.

oocytes and the subsequent return of an embryo, would all seem to be well justified by the possibility that a long-desired pregnancy might thus be established. "There would, of course, be the proviso that fully informed consent must first be obtained from both the patient and her husband, and *agreement reached that pregnancy would be terminated if monitoring revealed an abnormal fetus.*"

There the problems of what to do with a lab-made defective fetus caught up with Austin, despite the retransplant. As Edwards states: "The last thing *we* want is abnormal babies." [14] But what if the deformity was discovered only late in term or at birth? What would be the precedent-setting implications of terminating the fetus at this point?

Also, was it so obvious that prospective parents had to consent to such an act? Was it proper for a doctor to demand that a woman not have a defective child if she—now that she carried it—wanted to have it? And what if she could not have another one because she was infertile and the Austin way had not yet been perfected? Was it for the doctor to judge whether she could have no child of her own? How abnormal a child would the doctors disqualify to live—one with an 80 percent, 50 percent, 10 percent disability? And was the purpose of the agreement to protect the parents—or the doctors, who, at this experimental stage, feared the outcry, "breeders of Frankensteins!" an outcry that could stop the flow of funds from sources sensitive to public pressure or even lead to prohibition of such studies?

At this point Austin turned for the first time to the ethical issue implicitly raised by his frequent inspection of the fetus to see if it was normal. "The absolutist stance (on the basis that the conceptus [or fetus] has inviolate rights) is logically indefensible, for it depends upon the arbitrary identification of a particular stage—commonly fertilization—as the start of individuality, the moment when, according to informed theology, the 'soul' is instilled. The view is illogical because it assumes that the oocyte lacks the potential for development until fertilization occurs; the mammalian oocyte can, in fact, enter upon embryonic development parthenogenetically, and even in the very limited studies that have been made with experimental animals, progress has been recorded as far as the late fetus. To be sure, the developmental machinery in the oocyte normally waits upon sperm penetration, but this is not invariable. Apart from supplying the usual signal for development to begin, the role of the spermatozoon is essentially genetic—its chromosomes make possible the birth of males and carry the paternal genes.

"The commonsense assessment must therefore apportion very minor rights to the preimplantation embryo, certainly no more than to an ovary with its many thousands of oocytes. On these grounds the initiation of embryos *in vitro* and their use in the course of attempts to meet genuine human needs are acts of equal ethical standing to those acceptable for medical therapeutics.

"Some concern may be felt that routine work on human embryos, involving inevitable large losses through technical deficiency and the diversion of many for purely informational ends, may tend to erode the standards on which a proper regard for human life rests. But this surely is special pleading in a society that has learned to live with abortion on request. Indeed it is arguable whether even this practice really strikes at the foundations of ethics." In conclusion, Austin said: "Tacitly, if not explicitly, the futility of absolutist philosophy and the necessity for unending compromise in human affairs are beyond doubt accepted as the principles governing human behavior."

The deep silence attending Austin's paper lasted a good while after he had finished. Then, as if by a hidden signal, more hands were raised to indicate requests for the floor than had ever been raised before in the two days of meetings. The discussion that followed reflected the complexity of the issue; it was not very focused, but it did touch on many of the aspects involved.

First, the question of the rights and status of the fetus was raised. "I just wanted to note that Austin discusses the fetus as if it were comparable to a *cyst,*" stated Prof. Jérôme Lejeune, who is with the Institut de Progénèse in Paris, and is one of the world's distinguished geneticists. Lejeune, who discovered the detectable chromosome abnormality in mongoloids, is also a leading Catholic. Now he delivered a fiery condemnation of the aborting of genetically deformed fetuses, which he repeatedly characterized as "children." He was particularly upset by the fact that some physicians perform experiments on fetuses about to be aborted. Like many true believers, who come to meetings primarily to make their own opinions felt, he unrolled his banner whenever the opportunity arose.

"The rights of the fetus and embryo must be logically regarded as a progressive process," Austin answered. "You see, according to certain philosophies, the fetus, or the embryo, acquires full human rights right from the beginning, the beginning being in that case, fertilization. This raises difficulties with regard to the procedures that are involved with *in vitro* fertilization. But I think most people are prepared to accede these days that the logical viewpoint is to accord gradual progressive rights to the embryo from the beginning onwards, and accord it full rights only at birth or even, some say, afterwards. That part I do not want to debate, but it is the question of acquiring progressive rights that is important."

I wondered if Professor Austin realized that he was defining his beliefs as "logical," and by implication, those he opposed as "illogical." Actually, both positions are matters of value judgment and subject to changing social mores. There is no *scientific* reason to accord the value-loaded term "life" at a specific point; an ideological case can be made for many points. What can be rationally discussed is the social and personal consequences of various definitions. One extreme definition makes abortion a crime; and masturbation and *coitus interruptus*, sins (because these practices deliberately "kill" sperm). The Bible says: "It was evil in the eyes of the Lord" to "spill it [the sperm] on the ground," which some rabbis interpret as a condemnation of masturbation and others as a taboo against "threshing within and winnowing without."

At the other extreme is Austin's position, according to which "gradual progressive rights" are assigned to the embryo "from the beginning, and ... full rights only at birth or even, some say, afterwards." If life is defined as beginning shortly *after* birth, then doctors can kill a child born severely deformed without their action being legally or morally defined as "killing," because the child is not "alive." This practice occurs when a badly malformed newborn—e.g., one with a congenital heart defect or intestinal obstruction—is not given available surgical help [15] or is "put to sleep." [16]

Norman Podhoretz reports that in 1972, in Washington, he attended a conference of geneticists, biologists, medical men, writers, philosophers, and theologians. He found "a good deal of quiet support at the conference for what is called 'negative euthanasia'—that is, refraining from medical or surgical procedures which might be necessary to keep a mongoloid infant alive and allowing it to die instead." As to "positive euthanasia," he reports that though it "had no public partisans at the conference, it did have a few private defenders." [17] One molecular biologist "of the greatest renown" suggested that "no newborn infant shall be declared human until it has passed certain tests regarding its genetic endowment." [18]

I was wondering: Is the refusal to use newly perfected surgical techniques on a monogoloid child with a blocked intestinal tract to be defined as "killing"? Should parents be given this second chance at abortion?

Should the Declaration of General and Special Rights of the Mentally Retarded (which was adopted by the Assembly of the International League of Societies for the Mentally Handicapped in 1968) apply without exception? To infants one day old? This declaration states that a mentally retarded person "has a right to proper medical care ... no matter how severe his degree of disability."

For Austin, more was at stake than the question of whose belief might prevail in defining the status of the fetus. For him, a definition of the fetus as alive would prevent his experimentation, which entails terminating the "life" of many fertilized eggs; some decompose before they are ready to be replanted; some implants do not take; and in some replanted fetuses abnormalities arise, and Austin would prefer to terminate them.

Lejeune, not easily put down or kept down, raised his moral flag to wave in the face of Austin's "logic." He observed wryly, "Austin answered my question about the rights of men from a biological point of view, but the issue seems to me to be a moral, and not a biological, one."

Next, with Lamy's question about how much progress had actually been achieved, the discussion veered in a technical direction.

Professor Austin replied: "The most extensive work on fertilization *in vitro* involved the transfer of the subsequent cleavage embryo to a recipient female, and the ultimate birth of young, which has been done in mice. And these provide the best series. There is some work in rabbits and there is some work in hamsters. But most of the animal observations relate to embryos that are recovered from one source, from one animal—namely, the cleavage stages—and transferred to other recipients. But work is in progress at present on the fertilization *in vitro* of sheep and cow eggs. Some success has been obtained, some cleavage embryos have been observed. The next step is underway; it is a question of technical detail and I have no doubt that in due course we will see this procedure perfected in the bovidae as it has been in the mice and hamsters."

His statement concurred with other sources I had read which indicated that the process had technical difficulties but none of a major kind, and that there was no reason or theory why the procedure could not be carried out on people. Several reputable scientists have predicted that by 1995 eggs would be fertilized *in vitro* and implanted in surrogate mothers. [19]   Others expect it earlier. Dr. Watson writes: "We must assume that techniques for the *in vitro* manipulation of human eggs are likely to become general medical practice, capable of routine performance in many major countries, within some ten to twenty years." [20]

Dr. E. S. E. Hafez, an internationally respected American biologist, was reported as commenting on the basis of his own work on reproduction, that "within a mere ten to fifteen years a woman will be able to buy a tiny frozen embryo, take it to the doctor, have it implanted in her uterus, carry it for nine months, and then give birth to it as though it had been conceived in her own body. The embryo would, in effect, be sold with a guarantee that the resultant baby would be free of genetic defect. The purchaser would also be told in advance the color of the baby's eyes and hair, its sex, its probable size at maturity, and its probable IQ." [21]

As to cloning, "it has already been done in amphibia," says Lederberg, "and somebody may be doing it right now with mammals. It wouldn't surprise me if it comes out any day now. When someone will have the courage to try it in a man, I haven't the foggiest idea. But I put the time scale on that anywhere from zero to fifteen years from now. Within fifteen years." [22]

It all sounded like science fiction, but I wondered how many people listening in on or reading about these proceedings would realize how close scientists may be to these breakthroughs. Experiments on little mice may not seem very persuasive, although they do tell us much about human beings. But if it can be done to mice and cows. . . .

Until recently, the work on cows didn't get very far. The first calf to result from a transplant was born in 1950, and until almost twenty years later, it was still necessary to kill the mother cow to extract the eggs. Only in the past five years has it become possible to extract the eggs from the donor mother in a minor operation. An Oklahoma rancher and a Texas veterinarian have teamed up to sell eggs from high-quality cows; the eggs produce "purebred" animals which weigh hundreds of pounds more than their inferior host mothers and have more "saleable" meat.[23] In this way "donor" cows can be made to yield dozens of offspring a year. (For your classy fertilized eggs, write to Livestock Breeders International, Inc. Enclose a check and a form requesting second thoughts before the procedure is applied to humans.)*

The time to reflect on these matters is now, before the new techniques are applied to women, gain wide acceptance, and therefore become quite irreversible. The necessary reflection will certainly require a greater effort than can be made in a three-day meeting in Paris.

Gellhorn commented on this very point from his presiding chair: "I believe there is—I sense it, anyway—a consensus developing within the conference for a recommendation that there be some sort of a body of biomedical scientists, social scientists, environmentalists, and others who would join in the consideration of some of the issues that have been raised today and yesterday. And therefore, as I listened to the three presentations, I was just thinking in my own mind how such a body would formulate the questions that might be put to Austin, whose presentation I found very interesting indeed. Again it seems to me that already a number of questions have been raised with regard to this, but I would wonder again if there were another body listening to it, that at least one of the questions that would come up would be: What is the *priority that one would allocate to this type of study as compared with others*? And I should imagine if one is thinking of the allocation of resources, that in this instance the question would come up as to the social relevance of a particular experiment. Recognizing all of the difficulties that are involved in such a judgment, nevertheless, I would wonder if this would not come up as a very prominent sort of issue."

Gellhorn was again implying the need for a review mechanism, my pet Health-Ethics Commission, which warmed my heart. He also

*As these lines go to press it is reported that a cow gave birth to a calf grown from another cow's embryo, which was frozen for a week before it was implanted. *New York Times,* June 8, 1973.

raised an issue which had not been previously discussed but which was quite urgent: the question of allocation of medical resources. I assume Gellhorn meant that if one were to accord low priority to overcoming infertility—as compared to, say, cancer research—there would be no reason to stop these studies, certainly not as long as plentiful resources were available to subjects even less important. For instance, sizeable medical resources have been available for five hundred sex-change operations undertaken in the United States alone over a six-year period. [24] And God knows how many plastic-surgery operations have been done, from rhinoplasty (nose job) to blepharoplasty (fixing eyelids) to rhytidectomy (wrinkle removal) to restoration of virginity. If the ethical evaluation of Austin's handi-work had heretofore focused on its intrinsic value, that is, whether it was right or wrong by some abstract criteria or value—now issues of consequences were raised. Maybe his work was quite tolerable by one's definition of life but not in terms of the amount of personal or public good it produced in a world riddled with need. Nevertheless I felt that such considerations could not be used to outlaw the work. And one could hardly ask for a reduction in the priority given to the work, since it is—quite appropriately—rather low already. Finally, as long as we do not order other scientists to drop topics of limited social value, it is unfair to use social relevance as a reason for terminating this line of research.

Fraser brought the discussion back to a moral question—not concerning the status of the fetus, but that of the mother. "I would like to ask Professor Austin a question which has been asked before in the literature and to which I would like to hear some kind of answer. Do these women know that the implantation will not be made in their own case?"

On this point, Kass has written:

Who are these women and how did they come to "volunteer"? In the report describing the first successful fertilization and cleavage of human eggs obtained via laparoscopy, there are only several passing references to "patients," and the one-sentence abstract of the paper only increases our confusion by its use of the word "mother": "Human oocytes have been taken from the mother before ovulation, fertilized in vitro, and grown in vitro to the eight- or sixteen-celled stage in various media." If the women were indeed patients for infertility, then "the mother" is surely the one thing that they are not. In the recent Scientific American article by Edwards and Fowler, there is this solitary comment: "Our patients were childless couples who hoped our research might enable them to have children." We are not told, and can therefore only guess, as to what these women were in fact told. From the report that the women and their husbands had hopes, we can surmise that they considered themselves to be *patients*. But for the present, they are *experimental subjects*. One wonders if they were told this. [25]

It seems to me that whether or not these women are called "experimental subjects" (to avoid the loaded term "guinea pigs"), they are surely not patients in the traditional sense. I could believe that some women might volunteer to help science, but wondered whether the scientists weren't playing on the hopes of these would-be mothers. Edwards was quoted as saying: "We tell women with blocked oviducts, 'Your only hope of having a child is to help us. Then maybe we can help you.' " [26] Kass also cited the case of a patient, Mrs. Sylvia Allen, who submitted to the treatment only because she hoped the fertilized ovum could be implanted in her womb in the next two to six weeks, but implantation was never carried out [27] (presumably because the fertilized egg did not live long enough).

"The short answer to Dr. Fraser is yes, undoubtedly," Austin was now saying. "The whole situation is explained to them in considerable detail, so they are perfectly well aware what the chances are—that, in fact, the chances are very small. And this is made very clear to them." He added: "Many of the patients are nurses, wives of doctors, even doctors themselves; they are, on the whole, a fairly well-informed group of people."

This, I felt, was reassuring. While these people also had emotions that could color their judgment, I found it difficult to see how one could find any subjects more knowledgeable and better able to comprehend the information given to them than these. And surely their voluntary participation was more indirectly coerced than that of prison inmates often used in studies. [28]

Austin next responded to the question of priorities raised by Gellhorn: "I gave some figures at the beginning of the paper, and this adds up to the point that nearly one-third of infertile couples are found to need this kind of treatment. That is quite a large proportion of people. And therefore it seems justified to push ahead because this procedure would appear to be the only way around their difficulty until such methods had been developed for tuboplasty or transfer of tubes from one person to another or something of that kind, which is really far more remote." (Tuboplasty refers to the procedure used to repair or reconstruct obstructed Fallopian tubes, the cause of infertility in the women from whom the eggs are removed.)

The complexities kept proliferating. While Austin was, of course, quite right in saying that if a third of a population could benefit from a procedure, it is surely not trivial, or of low priority; however he avoided the other aspects of the question. For instance, is infertility an illness? Kass raised that question very adroitly: "Why would anyone want to provide new methods for making babies? A major reason given is that, in many instances, the 'old' method is not possible. . . . Physicians have a duty to treat infertility by whatever means only if patients have a right to have children by whatever means. But the 'right to procreate' is an ambiguous right, and certainly not an unqualified one." [29]

In short, not having a child, or having to resort to adoption, is not an illness like cancer or even influenza. Surely fertility enhances happiness (although as long as there are children up for adoption, Austin *et al.* may be viewed as satisfying a rather irrational quirk). True, some people *are* rather miserable because they cannot have children. A personal letter in a newspaper advice column captures their feelings:

> My husband and I have been happily married for 25 years, and he is the dearest thing on this earth to me. I have tried in every way to be a good wife, but I have failed him in the most important way of all. For some mysterious reason I have not been able to give him children. I was never able to get him to consider adoption. He hasn't complained, but I can see the hurt in his eyes when he sees his friends with their children and grandchildren, and my heart aches so. What can I do to make it up to him?
>
> —Hurting [30]

But most people *are* open to adoption. In a recent study of attitudes toward population issues (conducted in the New York metropolitan area), 63 percent said they would consider adopting a child, and an additional 11 percent said "maybe." [31] Most people I know who have adopted children are generally glad about it. They love their children as much as parents who are fortunate enough to be able to conceive.

Edwards, after agreeing that "the physical health of the parents does not demand that their infertility be cured" and "some couples may find adoption to be a satisfactory answer to the problem, which would also relieve social and personal problems of children without parents," has written: "Yet the desire to have children must be among the basic human instincts, and denying it can lead to considerable psychological and social difficulties." [32]

Well, I am not aware of such an instinct at all, if by instinct one means a natural, immutable urge. It is, though, a sentiment in our culture that is strong and difficult to alter, although by no means unchangeable. Recent trends, in fact, suggest that it is changing now.

Dr. Arthur G. Steinberg, of the Department of Biology at Case Western Reserve University, Cleveland, Ohio, now returned the dialogue to a technical matter by asking a question that reflected his position: "My question is based on profound ignorance. Is it feasible to remove the egg in women who have a blockage of the Fallopian tubes and reimplant it on the uterine side of the Fallopian tubes so that fertilization may take place in the classical manner?"

In this way, a whole list of moral and legal issues could be avoided. But Austin, who was certainly not to be outflanked on technical grounds, stated, "Well, the point is really, I suppose, that the fertilization doesn't normally take place in the uterine end of the Fallopian tube but rather in the ampulla, which is the ovarian end of the tube. It is just possible that it might occur in that section, but in these cases the pathology usually extends throughout the Fallopian tube and mostly these women have Fallopian tubes that cannot be blown through and are virtually useless."

The next question, by Professor Hilton A. Salhanick, F. L. Hisaw Professor of Reproductive Physiology at Harvard University and head of Harvard's Department of Population Sciences, demonstrated how deeply technical and moral issues intertwined. "I would like to ask Dr. Austin a question on the available information, including animal data, demonstrating that the product does not cause complications; in particular, what I am interested in is the intellectual component. In other words, I know that we cannot find mice that look and behave and have functional units that are appropriate, but the most sensitive test of all is what is the intellectual capacity of the product. Now has that been examined?"

This issue is important, because when eggs are removed and

replanted, the fetus is *handled*. This might "fracture" it in ways which will not show until, say, the child is a year or more old. This clearly indicates that many more animal experiments, including fairly lengthy follow-ups, would be essential before the procedure is applied to humans.

Professor Austin was very brief this time: "Not to my knowledge, no." Thus he did not seem to be fully honoring the promise of his and Edwards's group that the tests on animals will be "exhausted" before *in vitro* is tried on humans (reported, for instance, in *Saturday Review*, September 30, 1972).

There was no doubt as to the reason for the breach in the discussion which followed in the corridor after the session. One scientist observed, quite angrily, "Primates are more expensive than women volunteers." Dr. John Case commented, "These guys are after a 'first'; they will not report experiments in humans that fail until they have a successful one."

"Yes," I added, "but success will mean a child born without manifest deformities. There will be no assurance as to his or her normal development for quite a few years. Several experts have pointed out that the fetus might be 'traumatized' by the handling and its mental development retarded."

Case just nodded in agreement. It seemed to me that this situation was rather different from that of amniocentesis, in which similar concern about the long-range effects, especially on the IQ level, have been raised. Aside from the technical fact that Austin's intervention is more extensive, the purpose of the former technique is the prevention of a severe illness, for which one may wish to take risks. But to overcome infertility in an era in which we seek to curb population growth and in which children cry out for adoption, should one take the risk of producing a generation of lab-generated, man-made, retarded children?

On the way out, I recalled two other issues raised in writings about these experiments; one was mentioned only briefly during the deliberations, and the other was not explored at all. The first concerned the sanctity of life. [33] Such experiments, it has been argued, undermine the respect for the inviolability of a living human. Our civilization, it has been written, requires people to be treated as ends, not means or objects; as having a special moral status, not something to be sawed apart or dissected at will. If such experiments will change our attitudes to the inviolability of persons, the whole veneer of civilization may be endangered. War, violence, destruction of civil order will be undeserving beneficiaries.

A second related argument was that these experiments violate the family. It was argued that the family is the essential core institution of all societies; without it societies will not survive. The family is especially vital in our time as the last haven of warm personal relations. As Kass wrote:

> Some of the important virtues of the family are, nowadays, too often overlooked. The family is rapidly becoming the only institution in an increasingly impersonal world where each person is loved not for what he does or makes, but simply because he is. The family is also the institution where most of us, both as children and as parents, acquire a sense of continuity with the past and a sense of commitment to the future. Without the family, most of us would have little incentive to take an interest in anything after our own deaths. It would be a just irony if programs of cloning or laboratory-controlled reproduction to improve the genetic constitutions of future generations were to undermine the very institution which teaches us concern for the future. These observations suggest to me that the elimination of the family would weaken ties to the past and present, and would throw us even more on the mercy of an impersonal, lonely present. The burden of proof should fall upon those believing our humanness could survive even if the biological family does not. [34]

Austin *et al.* were accused of undermining the foundation of the community by encouraging sexual unions outside the husband-wife "covenant of fidelity." [35] Visions were even raised about assembly lines of test-tube babies replacing the whole institution.

Both issues seem to me relevant but also greatly overdramatized. The writers are not students of society; otherwise they would have noted that while our knowledge as to what sustains the sanctity of life and the institution of the family is far from complete, we do know that both are affected by numerous powerful factors, of which scientific experiments are one vector among many, and probably not a very important one. In view of the war in Vietnam; wanton bombing of civilian populations in cities; the Mai Lai killing of children, women, and the aged; the fact that most criminals go unpunished; the fact that thousands are killed on the highways each month because we do not build safer autos, or are willing to make people use trains; why should we bother with Austin's lab and his barely visible fetuses?

As to more developed fetuses, the answer may lie in prohibiting termination of a fetus in the lab under the same conditions in which it is prohibited in a woman, say, once the fetus is four-and-a-half months old, or when it is "viable" (which would also mean not keeping it artificially alive, even after this point, if it is deformed beyond viability). Most of all, Austin's work was being undertaken in the hope of enhancing life and increasing happiness—not to be destructive. Nor should the replanting work be judged by assembly lines of lab-born babies or the development of artificial wombs or cloning. Each of these has to be evaluated separately. While they are related, each requires its own scientific and technical development, and one might oppose one or the other and still not ban Austin. I realized some held that "one thing will lead to another," but I was quite sure that this was not necessarily so, and I knew I would have to sort out one of these days why I felt that way.

As to the family, it surely was in trouble, but the sources of its undoing included economic changes (e.g., more employed women), demographic changes (fewer children per family in the United States), a general disintegration of taboos, and at least a score of other factors. Even Kass admits that:

> The congregation of deliberate family wreckers includes persons eager to remove all restraints from human sexuality or to render obsolete the biological differences between the sexes, others who see the destruction of marriage as a needed step in limiting population growth, and yet others who find the modern nuclear family a stifling and harmful institution for education and child-rearing. I will not deny that the modern nuclear family shows signs of cracking under various pressures. It may have intrinsic limitations which make it seem, even at best, ill-fitted for modern technological society. But perhaps this should be viewed as a problem of modern technological society rather than of the family. We really ought to be less frivolous and journalistic in discussing such matters, and should keep in mind the essential question: Are we to accept as desirable the final solution which eliminates biological kinship from the foundation of social organization? Yes, laboratory and governmental alternatives could be devised for procreation and childbearing. But at what cost? [36]

I found it hard to see how, even if all infertile women were helped by Austin and company, the family system would be further undermined. On the contrary, the tensions infertility imposes on a family seem at least as high as those that Austin's work may impose on it. And it seems that most people are quite receptive to the idea of sons and daughters of artificial insemination, though occasionally it does raise a psychic problem for the infertile man, because the child is not of his "blood," an ignoble sentiment we need not cater to. [37] Most of all, I failed to see the grand symbolism. If the family survived several million instances of adultery a year, would several thousand test-tube matings undo it? If prostitution failed to separate the biological from the social element in the sex and love life of those who do not resort to their cold services, could Austin *et al.* undo it for us? Or was the criticism one of easy humanism, which, in the name of vague, general, somewhat sentimental "theories of man," simply disregarded basic human needs of women and men desperate to have a child and stubborn about it being of their own flesh and blood?

I did not feel ready to pass a final judgment, nor was I called upon to do so. But I felt that while many specific serious questions were raised (Why not do more experiments on animals? Were the women really volunteers? What shall we do with a lab-made fetus discovered to be severely retarded once it is nine months old?), these were questions involving the details of the procedure. The *principal* objection, to the whole "business," seemed less powerful. As far as I could see, neither civilization nor the family were being tested or violated.

Once in a while I asked myself whether we would not be better off if the whole thing had never happened, but this was obviously an idle thought. The work was progressing, and once science took a bite from the forbidden apple of knowledge it could hardly be stopped, certainly not unless there was much greater evidence of undesirable results. At the same time, scientists need constant watching to prevent the manipulation of persons subject to tests and an irresponsible rush to use humans, and to make some difficult decisions about the fate of deformed fetuses. This left the more general questions of genetic engineering, of which *in vitro* work was but a small part. The next session was to unfold the wider canvas. It was time for a welcome lunch break, which both foes and friends of *in vitro* happily undertook together.

CHAPTER THREE

Are We Debasing Our Genes?

The next session dealt with the question of whether our increased intervention in natural processes would foul them up. By preventing elimination of the genetically unfit by natural selection, will we foster a weak human race—too weak, perhaps, to survive? By interfering with and trying to improve our condition, we have brought our environment to the point of crisis; are we about to do the same with our species? Some distinguished authorities have raised that spectre about the new genetics. Thus Sir Julian Huxley, renowned author and biologist, has written:

> ... it is clear that the general quality of the world's population is not very high, is beginning to deteriorate, and should and could be improved. It is deteriorating, thanks to genetic defectives, who would otherwise have died, being kept alive, and thanks to the crop of new mutations due to fall-out. In modern man the direction of genetic evolution has started to change its sign from positive to negative, from advance to retreat: we must manage to put it back on its age-old course of positive improvement. [1]

On the same issue Bentley Glass has said:

> ... by surrounding ourselves with an ever more artificial environment, we unwittingly modify the rigor of natural selection in many ways. The price we must pay, in the end, for the mercies of medical care and surgical aid is a dysgenic [detrimental from a genetic viewpoint] increase in the frequencies of certain detrimental genes, the effects of which we have learned to ameliorate. ... No one, I think, would have it otherwise. Yet to

contemplate the man of tomorrow who must begin his day by adjusting his spectacles and his hearing aid, inserting his false teeth, taking an allergy injection in one arm and an insulin injection in the other, and topping off his preparations for life by taking a tranquilizing pill, is none too pleasant. To say the least, medical science steadily increases the load it must carry. [2]

The session, in which technicalities proliferated, touched on a number of related problems. Was man in fact capable of taking his biological nature into his own hands and safely modifying it? Or was he destined to act out the laws of nature—however cruel or arbitrary—as if nature's will were that of a forbidding and intractable God? And if our intervention had some undesirable effects, was the only alternative a hands-off return to nature? Or could we correct these side effects without giving up our human thrust?

Dr. Arthur G. Steinberg presented the first paper. A tall, distinguished-looking man, he began by saying that the genetic pool was gigantic. "The 'gene pool' is the totality of the genetic information encoded in the DNA of the species. It includes all the genetic loci (that is, all the genes) and all the alleles [the many functional and nonfunctional forms a gene may take] at each locus in existence within the species at a given time."

Steinberg went on to ask, in highly technical terms, how many genetic loci we mean when we say "all the genetic loci." He then said: "There is no precise answer. An estimate can be made from the amount of DNA in the nucleus of a cell. The human sperm contains about 2.5×10^{-12} grams of DNA. Using 620 as the molecular weight of a nucleotide pair, the number of pairs per human genome [the gene set of a person] is approximately 2×10^9. It is commonly stated that a gene on the average is composed of about a thousand nucleotide pairs—that is, enough to code for slightly more than three hundred amino acids. If so, the human genome has about two million genes. But this is only the beginning of the determination of the size of the human gene pool—it merely measures the number of loci. Each locus may exist in the species in a large number of allelic forms. How many alleles, in theory, may a locus have? The number is enormous."

Steinberg explained that the pool was vast: "The average mutation rate per gene per generation is about 10^{-5} or 10^{-6}. Even the conservative estimate of 10^{-7} or 10^{-8} would supply adequate opportunity for the species to accumulate, over time, a large number of mutations." This vastness suggests that it is difficult to dirty the pool by human intervention, and that other forces affect it much more.

Environmental factors—for example, radiation from natural sources—affect all of us all the time, and cause a large number of gene changes. Among these, Steinberg seemed to suggest, the effects of several hundreds of thousands of abortions undertaken for genetic purposes would barely be noted.

Steinberg, who intoned his statistics and technical terms as if reading from a Shakespearean text, now reached his main point: "Selection implies that genes or gene complexes have a different survival value—survival being defined in terms of ability to leave offspring, and not in terms of longevity. Relaxation of selection pressure, regardless of how it occurs, will lead to an increase in the frequency of whatever gene against which pressure is relaxed. This raises the questions of (a) how rapid will the increase be, and (b) what effect this may have on the well-being of the species. Thus the advances in medicine, which result in saving those who would have died for genetic reasons, are increasing the frequencies of these so-called deleterious genes. I refer to the rapidly expanding technique of amniocentesis, followed by abortion of the affected fetus. 'Expanding' is used here in the sense of an ever-increasing number of conditions which can be diagnosed prenatally and in the sense of being more widely and frequently used—at least in the U.S.A."

Without changing his tone, he placed the apex on his carefully constructed intellectual pyramid:

"Various investigators have estimated the potential effect that the use of the procedure may have on gene frequencies. They have used different assumptions and they have approached the question from different viewpoints. All have arrived at essentially the same conclusion, namely, *that the effect will be slight*. Let me illustrate this with some estimates I have made for recessive lethal genes, both autosomal and sex-linked."*

*Lethal genes produce, via disorders in the sex chromosomes or the autosomes (nonsex chromosomes), defective traits whose effects can be so severe as to cause death. The sex chromosomes determine specifically male or female traits. Males are typically represented as XY and females as XX. Different combinations like XYY, XXY (Klinefelter's syndrome), and XO (Turner's syndrome), are abnormal. Some genes are X-linked (that is, located on the X chromosomes), as are genes for color vision and blood-clotting ability. Disorders in the X chromosome, then, might produce red-green color blindness or hemophilia. Other traits are autosome-linked; that is, they are not located on the sex chromosome but on other chromosomes.

"I have assumed two patterns of reproduction for autosomal genes," Steinberg continued. "(A) couples reproduce until they have an affected child at birth order ≤s [before they have the desired "sum" of children] or, if an affected child is not born, until they have s children [the desired number], and (B) couples avail themselves of amniocentesis after the birth of an affected child and breed until they have s normal children."

Steinberg was using the logical rather than the empirical arm of science. Rather than presenting data on what couples actually did under the circumstances, he made two assumptions. One described the behavior of parents who made no use of, or were not able to use, the technique of amniocentesis to insure the subsequent birth of healthy children. Therefore, they would stop reproducing after the birth of an afflicted child, out of fear of having a second ill child. At the other end of the scale were those parents who, having had an afflicted child before they reached their desired number of children, continued to reproduce, being guaranteed a healthy child through the use of amniocentesis. Patterns of reproduction that deviated from these two extremes would decrease the effects of medical care on the gene pool.

Steinberg built his argument on these foundations. This is particularly appropriate when, as regards the question at hand, the data available on what people actually do under these circumstances are very limited, and because logical steps can encompass situations which have not yet arisen and hence are not subject to empirical examination (for instance, that time when genetic engineering is more widely welcome and hence more widely practiced).

Steinberg continued: "The number of generations required to double the frequency of the gene under these assumptions is shown in Table 1." One could hear pages being turned throughout the hall to find the table, which was at the end of Steinberg's paper. The "q" at the top of the table stood for gene frequency, that is, the extent to which the defective gene is common; "s," on the side of the table, stood for the desired number of children.

Steinberg explained the table and its implications for the effects of medical interventions on the gene pool. "It is clear from the table that the *gene frequency will be doubled only in the course of centuries*, by which time medical practice and other sociological factors will be vastly different from what they are now and the problem confronting the species will be very much changed."

TABLE 1

The Number of Generations Required to Double the Number of Heterozygotes

---

			q		
s	.1000	0500	.0100	.0010	.0001
2	20	35	156	1,517	15,127
3	17	29	127	1,254	12,388
4	14	25	110	1,070	10,675
5	13	22	99	962	9,648

I examined the table. It indicated that the number of generations it would take to "dirty the pool" to the extent of doubling the incidence of heterozygotes ranged—depending on the details of the assumptions about how many children affected couples had and the frequency of the original defect—from a "mere" thirteen generations (or 390 years) at worst (assuming five offspring per average family and a high .1 rate of gene frequency),* to a staggering 15,127 generations (if there were two children and the original rate of affliction was a very low .0001).

I could readily see why one would not worry about events so remote in time. Completely new therapy might, some far-off day, well be available; a corrective message might be sent to a defective gene via a virus, or a missing gene might be supplied through genetic surgery. What I could not understand was why abortions of affected fetuses would *dirty* the pool rather than help to "purify" it! Wouldn't abortions remove the illness-producing genes, thus leading to a race ever less prone to illness? And, anyhow, what was the alternative? To let the deformed children be born and hope they would not live long enough to reproduce? This seemed not only less humane, but also *less* genetically "safe" than abortions; some people with severe gene defects would end up reproducing. True, in either case, the net effect

---

*The rate of affected individuals would be the square of the gene frequency, or .01.

would be small, because of the enormity of the pool, but at first I could not see that the effect of abortions would be in the problematic direction Steinberg implied.

It was only when I reread the lines Steinberg had just delivered that I understood. If all or most of those parents who discover they have a defective fetus will decide to abort it and will try to have another child (while those who would give birth to a defective child would be less likely to try to have more children), there might be a problem. Since many genetic illnesses do not hit each offspring (e.g., sickle-cell anemia), the next fetus (or the one after that) may be normal and live to reproductive age—but with the capacity to pass on the latent defective gene to its offspring. Hence, if most afflicted fetuses—which, without intervention would not have reached reproductive age—are replaced by fetuses that are "normal" in all but their hidden, inactive, sick gene, the gene pool will only get dirtier.

Of course, not all or even most parents will go on to have a normal child with a latent defective gene, especially if the public is educated to the undesirable implications of such a choice. People bearing such genes could either refrain from having children, adopt a child, or rely on artificial insemination. Other steps could be taken to keep the pool clean. If public health authorities would urge parents to complete childbearing when the mother is young and the rate of genetic illness is therefore significantly lower, this could make up for all, or at least part, of the "deterioration" of the pool caused by the genetic interventions which increased the number of recessive, sick genes.

Steinberg now turned to evaluate the dangers of genetic intervention. "I have been asked to discuss desirable and undesirable genes, and thus far I have avoided doing so because I am at a loss to know how to define them from the species point of view. The sickle mutation of the hemoglobin β [beta] chain is certainly undesirable for the homozygote [an organism with identical pairs of genes with respect to a heredity character]. Yet it was probably important for the survival of the African populations living in the *malarial* regions. Is it of any value to populations living in nonmalarial regions? Probably not, judging by its distribution in endemic populations."

The key phrase here was "from the species point of view." Sickle-cell anemia is clearly devastating to the individual affected by it (as distinct from carrying it in a recessive manner), but there is no reason, Steinberg argued, to worry about the effects of reducing its frequency on the society. Who knows, anyhow, what a species' needs are? Thus, while the sickle cell may be unnecessary for survival in New York City, it might be useful in malaria-infested Vietnam. Steinberg seemed anxious to support interventions on behalf of individuals and to reject opposition based on fears of the effect such interventions might have on natural selections and the "quality" of the race. Some, like Muller and Huxley, were worried about "dirtying" the race, but even they could not deny help to a sickle-cell mother now, for the future of the race.

Steinberg continued: "Similarly we can agree that the homozygous condition for the gene leading to cystic fibrosis is undesirable (although less so now than twenty years ago) for the affected individual. But is the allele undesirable for the Caucasoid race in which it is so frequent? It must have been advantageous to the race at least some time during the race's evolution. It may still be advantageous, but we have no evidence to show this." (The victims of CF, or cystic fibrosis, suffer from blocked respiratory passages and pancreatic ducts caused by the excess secretion of a viscous mucus so excessive that it threatens to choke the patient. Even following treatment, the patients suffer from malnutrition, diarrhea, and breathing difficulties.) Did this mean that the CF-causing gene should be kept alive, "just in case"?

I wanted to be sure to hear every word of the rest of Steinberg's paper, so although he was reading clear English, I put on my earphones to amplify his voice.

"My point is that we know very little about the value of a gene to a given race or to the species. We know only about its value to the individual carrying it, and then only in instances where the effect is severe. In the light of such ignorance, it seems to me that the best procedure is to avoid all changes in the environment which are likely to change the mutation rate and to concern ourselves with alleviating the suffering of affected individuals and of those who may have affected children."

I later discovered that whether we want to or not, we are not now capable of eliminating CF, but we can and do treat affected individuals and even bring them to the age at which they will reproduce. Treatment thus might increase the frequency of CF genes; but Steinberg had explained that the effect on the gene pool would be small, at least for the coming several hundred years.

Steinberg continued: "In closing, I remind you that the quality of a gene or genotype may be determined only by the reaction of the associated phenotype in the environment in which it exists. A phenotype may be disadvantageous in some environments, essentially neutral in others, and advantageous in others. In the face of a rapidly changing and entirely new environment (new in an evolutionary sense), I do not believe that we can determine the value of specific genotypes to the species."

During the coffee break which follows the presentation of scientific papers as surely as spring follows the winter, I turned to my friend, a French sociologist who had dropped in for the session, and said: "I wonder what Steinberg would say if one reminded him that the society is made of nothing but a lot of individuals."

My colleague seemed puzzled. "So?"

"Well, the argument that you can't tell if a gene is 'good' or 'bad' has meaning to individuals as it has to society. Sure, I see that Steinberg tries to de-couple the two so that we could help individuals without worrying about societal consequences or have to act in the name of those. But wouldn't individuals also fear bringing up 'weak,' 'unfit' children? And would they not worry about the breakdown of civilization?"

"I see your point," my colleague responded. "And conservative doctors will use it to bolster their reluctance to use the new techniques to service individuals."

By now the audience was drifting back into the hall to hear the next paper. Maybe this one presented by my new friend, G. R. Fraser, would alleviate my concern.

Fraser's manner was rather different. He was younger than Steinberg, and less known in the field, although his credentials were impressive (M.D., Ph.D., Professor of Human Genetics at the State University of Leiden, Holland). His starting point, though, was the same as Steinberg's:

"Much concern has been expressed about the deterioration of the genetic endowment of the human race . . ." he read in a low, even voice. "The reasoning underlying such concern may be illustrated by a few simple examples. Retinoblastoma is a malignant tumor which, if untreated, is almost invariably lethal, and even if treated, usually leads to loss of sight. Bilateral retinoblastoma is very often, if not always, transmitted in a Mendelian autosomal dominant manner. Thus treatment which increases the survival, and hence the fertility or relative fitness, of affected individuals leads to a corresponding increase in the incidence of the mutant allele."

This was a case in which, because the illness is carried in a dominant manner, abortion of the affected fetus would help, however slightly, to purify the pool rather than dirty it, though of course, detection of dominant diseases of the fetus is still rather difficult. There would be no replacements in the form of healthy, reproductive carriers of recessive genes.

Fraser was extending his argument: "In the case of an autosomal recessive disease such as phenylketonuria [PKU], recent advances in understanding the effects of the mutant allele and in their alleviation by dietary treatment has again led to increases in the fertility of affected persons."

It would thus follow that not only amniocentesis and therapeutic abortions, but also dietary treatments and other medical innovations (insulin shots for diabetics), lead to dirtying the pool. I could see no difference in principle between keeping a PKU gene in the reproductive cycle by preventing severe retardation through a diet, and amniocentesis and abortion for CF.

And this was where Fraser was indeed headed:

"Even when techniques of prevention such as ante-natal diagnosis and selective abortion of affected fetuses are applied in the case of an autosomal recessive disease, there is a potentially dysgenic effect in that there is a tendency to replace the aborted fetus, whose chance of reproduction may have been very low. The replacement is by a child with normal viability and fertility, who has a two-thirds chance of being a heterozygote. Thus the number of mutant alleles transmitted to future generations is increased. This dysgenic effect will, however, be moderated by three factors. First, reproductive compensation may occur to some extent, even in the absence of selective abortion. Secondly, the identification of a marriage between heterozygotes is at the present time usually dependent on the birth of an affected child; counseling therefore, may be termed retrospective in this situation, and any reproductive compensation will not involve the first affected child but only those conceived subsequently."

In plain English, Fraser was saying that the undesirable consequences of treatment for the pool were "moderated" if the disease was "recessive" and hence recognized only after the parents had already had one afflicted child! Was this like saying that the costs of repairing automobiles would be reduced if they were repaired only after accidents occurred, instead of recalling all those suspected of a fault for tests and preventive corrections?

Fraser continued:

"Thirdly, the extent to which selective abortion is applied will depend on the age of the first affected child when the autosomal recessive condition in question can be recognized and will be maximal when this can be done at birth. At the other extreme, if the condition cannot be recognized till seven years of age or later, applicability of selective abortion will be minimal since many couples will already have terminated reproduction by the time the diagnosis is made in the first affected child."

The fact that some recessive illnesses are detectable only when the person reaches advanced age (e.g., some forms of blindness), after their similarly afflicted siblings are already born, seemed to me to favor the preventive methods—even if they muddy the pool—over acceptance of the illness. Under these circumstances, post-hoc genetic counseling and attempts to limit the number of offspring to be sought by the afflicted families would be even less effective in curbing the genetic curse than they would be if the illness were recognized when the first child is born. If one waits here for illness to exhibit itself in the children, it is too late for all, or most, family planning.

Fraser spoke as if he had read my mind: "The situation would be quite different if *premarital screening programs* are implemented for heterozygosity at a number of loci where autosomal recessive conditions are determined; this introduces the possibility of prospective counseling. Premarital detection of heterozygotes is already feasible on a wide scale in the case of a disease such as sickle-cell anemia, and it will become so in the future in an increasing number of other recessive conditions, both autosomal and sex-linked. As mentioned above, this will lead to the possibility of giving genetic advice to couples before, rather than after, the birth of an affected child [prospective rather than retrospective counseling]."

Maybe one day all couples, or at least most, will find it advisable, even fashionable, to have their genes typed before they plan their family, even before they marry. Thus, if they have a high chance of having a mongoloid child but do not wish to use abortion, they may decide, because of the high costs involved in bringing up a mentally defective child, not to have children at all, to adopt some, or to have one less. Fraser was now changing course: "So far I have mentioned only potentially dysgenic aspects of prevention and treatment. These will be balanced by other advances in medicine and in our understanding of human biology, which will have opposite effects. Thus genetic counseling in the case of recessive disease, whether prospective or retrospective, may be followed by a decision to abstain from reproduction rather than by antenatal diagnosis and selective abortion; this will clearly have favorable effects in reducing the number of deleterious alleles. In the case of dominant disease also, a better understanding of the basis of transmission may influence affected persons not to reproduce, even though their potential fertility may be improved by treatment."

This surely proved to be the most difficult session; the terms were unfamiliar, the issues complex and technical. I was looking forward to the next session, in which social and ethical issues would again be at the center. But I also realized that if nonscientists wish to deal in the consequences of science, they had to make the effort to acquire enough knowledge to be able to follow the main findings and propositions. Fraser was now reading:

"Artificial insemination by donors is another form of prevention which need not have dysgenic effects, and in some situations, could actually have favorable effects in restricting the dissemination of deleterious alleles. Thus when the male partner in a marriage is affected with a dominant or sex-linked recessive condition, this technique provides an excellent method of arresting the spread of the mutant allele. In the case of an autosomal recessive condition, such a technique is also probably not dysgenic, since abstention from reproduction by the male partner of a couple which adopts such a solution—whether counseling is prospective or retrospective—will reduce dissemination in the next generation of that particular allele, though of course use of donor sperm may increase dissemination of other deleterious alleles. When screening for heterozygosity for a wide variety of alleles causing autosomal recessive conditions is possible, it would be wise perhaps to restrict sperm donation in such situations to males who are free of those which can be detected, though of course this will always be only a subset of all such alleles."

I could see husbands who are afflicted with some terrible heredi-tary illness agreeing with their wives on artificial insemination for the sake of their own children in the case of dominant disease, and for the sake of their grandsons in the case of a sex-linked disease such as hemophilia. But if they were carriers of a recessive disease, how many would agree to it for the sake, not of their children who are unlikely to be afflicted, but for the sake of the anonymous worldwide genetic pool? Could one convince them? Should one try? These were matters I decided to earmark for future deliberations.

Fraser meanwhile turned to an interesting possibility. "Probably outweighing all these factors in the control of disease due to chromosomal aberrations is the strongly eugenic effect of age changes in reproductive patterns." He argued: "Strong associations between increasing maternal age and the incidence of certain chromosomal aberrations in the offspring have been noted. Clearly, therefore, the marked trend toward lowering of parental ages, in economically developed countries at least, has been having, and will continue to have, very favorable effects on these incidences."

Concern, then, with the effects of individual preventive actions on the societal pool was uncalled for, not so much because the effects are minimal (their size loomed or shriveled depending on the scale you used, whether gene statistics or depth and scope of individual sufferings), but because many other shifting factors could more than make up for whatever damage therapeutic abortions and other genetic measures on behalf of individuals might cause to the pool. This would be the case even if new interventions to help individuals were practiced on a very wide scale and if many parents were to replace the aborted fetus with normal children carrying those sick genes recessively.

Actually, it seemed to me, in the rush to promote birth control, we did not emphasize as much as we could have that the best of all possible worlds—both for the family and for society—was to be achieved not only by having fewer children but also by having them earlier. I noticed my thoughts were wandering, but I felt it was a subject worth pursuing. The unwitting damage caused by the anti-population propaganda, which advocates delayed parenthood, was a typical case in point where the concern with one dimension—overpopulation—was not balanced by attention to another, here, genetic health. Policy makers and the public find it easier to deal with one dimension at a time than to deal with multidimensional problems, or to identify cures which would serve two ends: fighting both pollution and poverty (e.g., certain public works); both unemployment and absence of early education (day-care centers run by unemployed mothers), etc. In this case, limiting family size and doing so at an early age would serve to improve living conditions as well as genetic health.

Fraser was now summing up his view of the situation at present. Like Steinberg, he was not using actual empirical data on family size and on what parents actually do following abortions of afflicted fetuses (many seem not to replace them),[3] and instead he used logic to conclude that, even if we make the most unfavorable assumptions about "dysgenic" effects on the pool, we need not hesitate to follow whatever medical and public policy may be attractive on other grounds.

Next Fraser confronted what I came to view as the Red Herring Number One of the field. This was a kind of *deus ex machina* which descended when those who held that we need not worry about purifying the pool ran out of all other arguments:

". . . Another argument frequently adduced by the prophets of genetic disaster is that, by favoring the propagation of deleterious alleles, medicine is reducing the chances of survival of mankind after a general catastrophe such as nuclear war, since the genetic endowment of alleles advantageous in such a situation will be seriously depleted. In the case of diabetes, this argument has been confuted by a suggestion of Dr. J. V. Neel that diabetes may represent a thrifty genotype from the point of view of carbohydrate metabolism, which was of advantage in the past when food supplies were very limited and is only deleterious in the context of the grossly excessive and unbalanced diets which characterize our civilization. If this is true, and it is at least as likely to be true as the foreboding of the pessimists, then increasing the fertility of persons with this genotype, either by preventive or therapeutic advances, is actually eugenic insofar as a reversion is feared, after a catastrophe such as a nuclear holocaust, to the circumstances which prevailed before the introduction of our technological civilization."

Fraser thus endeavored to show that if the societal value of a gene was unknown or ambiguous, increasing its frequency could turn out to be as useful as it could be harmful. In my book, if the point is "We don't know," this is as much a case for inaction as it is for action.

Fraser himself concluded: "I personally feel that methods of prevention and treatment of genetically determined disease which are available today and which may become available in the near future should continue to be applied in the context of the individual family unit, who should decide how to act, taking into account the acceptability of the appropriate technique in the light of their own personal religious, moral, social, and economic circumstances and, insofar as possible, the interests and rights of their unborn or unconceived child."

In his closing remarks Fraser said: "It does not seem to me that application of these methods should be affected by hypothetical societal goals of which a universally acceptable definition cannot be provided and whose potential benefits cannot be accurately predicted because of inadequate knowledge and insight. I do not believe that continued application of these methods on the present scale, or even on the substantially extended scale which may be introduced in the future, will seriously jeopardize the genetic future of mankind. This is not to say that very serious problems which concern human reproduction do not exist, but these are primarily quantitative rather than qualitative. The methods of prevention and treatment of genetically determined disease under discussion should perhaps be regarded as ancillary to the educational, biological, and medical armamentarium which is necessary to avoid the catastrophe of gross overpopulation of our planet which would render all discussions of this type entirely nugatory and irrelevant."

Though Fraser's route was somewhat different from Steinberg's, the conclusion was identical: we could service individuals; we need not worry about the societal pool; it was not wise to act to improve the pool. I was left uneasy by this complete de-coupling of individual and societal treatment; nor was I sure if we should not act for societal goals. But for now—I did not quite know why I felt this way; I just felt I should think more about all this.

During the discussion, I asked a question on a different matter: "Dr. Fraser, how effective is genetic counseling? I read a study conducted at Johns Hopkins which shows that many patients ignored the advice given. It is not so much that I would give up on this preventive technique, but one may have to work more with the patients until they can accept and use the information. And may I ask what would be the effect of preventive abortions on the couples' grandchildren and great grandchildren?"

Fraser replied: "As regards the problems of counseling, it is, of course, self-evident that at the moment counseling involves a very small, largely self-selected proportion of our population. I'm a little more hopeful than the experiences of Professor Etzioni's colleagues has made them, because of the very fact of self-selection, since people come to a counselor when they are worried, and for this reason are likely to be receptive to advice. However, it does not surprise me at all that a proportion of people do not take account of the advice, or perhaps I should say the information, because it is not very clear to me how anyone can advise them. This is true even of the simplest case which may not necessarily be treated by a genetic counselor at all, such as the repetition of catastrophe with which the parents are familiar because it has already occurred in a previous child. When the risk of such repetition is twenty-five percent, there may well be circumstances where the parents for one reason or another will wish to take this risk, and I don't see any kind of coercive or directive possibility of advising them to the contrary. There may be some who take the risk, not deliberately, but through negligence of one sort or another, and I would hope that this proportion will decrease. I would welcome any kind of mechanism whereby these catastrophes by negligence can be averted, whether it is by referral to a psychiatrist or to a gynecologist or any other specialist who would like to participate in such a process. It is really a question of education and progress, and even in the countries which we like to regard as developed, it is unfortunately true that a large majority of the population is not sufficiently intellectually endowed, or perhaps motivated, to take advantage of the services which exist. I think this is a great tragedy because it seems to me that to have a child at all is a great responsibility and that all possible medical advice should be very carefully sought and evaluated with respect to every pregnancy.

"As regards the point Professor Etzioni made about the impact of counseling on future generations descended from the patients whom we are counseling, I think, in view of our own uncertainty, and in view of the uncertainty of people who should know most about this field, it's unfair to expect the patient to become involved in the future of the human race in the abstract. I think some of them will become involved in the future of their grandchildren, but I think going beyond that is a little difficult for them."

Professor Lamy, with whom I had lunched the first day of the conference, raised the issue of responsible and compassionate counseling: "Who is to give this counseling? Certainly it cannot be given by a geneticist whom I will call 'pure,' that is to say, without clinical knowledge. Certainly this genetic counseling should be given by a doctor, in a hospital, and the diagnosis must be formulated in a precise way and with the help of specialists. Obviously, if the diagnosis is inexact, it is bad counseling.

"The second point is the way in which genetic counseling should be administered and how explanations should be given. Certainly every effort should be made to relieve parents of any guilt feeling, by explaining to them that they are not guilty or responsible but just victims of an unfortunate hazard. Also, things must be explained with a great deal of precision. Many parents do not realize what 'one in two chances' or 'one in four' means. I have the habit, at the Children's Hospital, of using a roulette wheel or a deck of cards to explain both the alternatives and the series. Many people are completely closed to this kind of thinking. Knowing to what extent our advice is heeded would make our medical counseling worthwhile. It is certain that counseling should not be given only once, but that the family should be revisited and the counseling repeated.

"Since at this meeting there seems to be a question of ethics, I must point out certain situations that can arise from people having knowledge of the facts. I can cite two examples. One day a man came to see me. He had had with his wife two myopathic children, with the particular kind of myopathy called 'Duchenne muscular dystrophy' (that is, sex-linked), and he said, 'Since I have one in four chances of having another child with the same disease, I intend to divorce my wife and have normal children with a different partner.' Another man who had had two hemophilic boys said to me: 'If I understand correctly, my wife is what you call a carrier of hemophilia. So if I have a son, he has one out of two chances of being a hemophiliac. And if I have a daughter, she also has one chance out of two of being a carrier like her mother. I am going to change wives.' I am by no means making a moral judgment, but simply wish to point out certain dangers. I do not suggest that we must not inform the families; but, as doctors, we must treat this question with moderation, prudence, and wisdom."

Moltmann next returned the topic to Fraser's central theme by asking about the general purposes for which genetic intervention may be used:

"Mr. Chairman, there seems to be an ethical judgment in the paper of Dr. Fraser which is very important, and as I understand it, can be generalized. He said that the prevention and treatment of genetically determined diseases is ethically justified; but then he hesitated to defend programs dealing with a more generalized control of the quality of human reproduction. I think there is an ethical consensus that genetic disease should be prevented and treated and corrected if possible, because there is a consensus in society, and even in mankind, about the evil character of these diseases. But this is true, perhaps, for one percent of the population—those with mongolism and other diseases. On the other hand, there can be no ethical consensus, and not even the political means, for one hundred percent control of all human reproduction. Therefore programs for improving the human race to create a eugenic paradise, with geniuses like Aristotle and Plato, or humorists like Charlie Chaplin, is an impossible dream, and is no ethical question at all. So, *summa summarum*, there can be an ethical principle to work like this: There is a consensus about the negation of the negatives; but there is no consensus about the positavatum of the positive or the improvement of a positive. So we have to work to overcome the commonly acknowledged diseases, at least for the present, and not to try to slip out of the present into the future."

Steinberg, who I knew had vast experience in these matters, seemed to react to my previous comment:

"The implied point of view, about the purpose of genetic counseling, is one with which I don't entirely agree. The implication is that the purpose of genetic counseling is to convince the counselees that they should have no more children if there is an unsatisfactory genetic disease in the family. I think the purpose of genetic counseling is to relieve the counselees of the worries that brought them to the clinic; to give them what information you can give them; and if, after they have received this information and you have made valiant efforts to get them to understand, they go on to have children, I think that decision is a correct one for them, and I think that the counseling has served its purpose—it has relieved them of the worries that have brought them to the clinic. This is the only purpose of genetic counseling."

This is a position doctors frequently take, because they are both ideologically individualistic and laissez-faire oriented, and because they are trained to worry about their patients, not about families or children of patients. The society at large takes a backseat. This position greatly simplifies decision making because all other, often conflicting, criteria are ignored. But is it the wisest course to follow? How could other considerations be brought to bear without removing the priority of the individual? Was it all right, for instance, for society to agitate for certain measures (the way it does for smaller families), as long as the pressure did not turn into economic or legal coercion?

Later, when I discussed the matter with Fraser in private, he offered the following analogy: "If we prevent a heart attack, the same questions may be raised: What are the consequences for society? The costs? Even the dysgenic effects? That is what medicine is all about; it would collapse if we would worry about society every time we treat an individual."

"I can see that," I thought aloud. "But don't we also occasionally ask these kind of questions, and have to do so more often as the pressure on societal medical resources rises? Doesn't refusal to think about it basically mean that the very rich get all the services they can use, but no one else?"

Mme Marie-Pierre Herzog, Director of UNESCO's Division of Philosophy in Paris, spoke of the need of the modern parent, and especially the parent faced with the spectre of inherited disease, to reevaluate traditional attitudes toward child-rearing and "natural," "blood" children:

"I want, in effect, to underline the need for a real theory of substituting satisfactions which would maintain or help maintain all those who are in a position of adopting a decision which might seem mutilating—for example, not having any children—in the name of scientific rationale or the survival of society, or even mankind. Substituting satisfactions are important, because one must not demand from a human being the renunciation of certain things without offering him other alternatives. In my view there are cases where substitutions cannot be found or else it is very difficult—drugs, for example. On the other hand, when it comes to natural children there are substituting satisfactions which have been practiced in the past by mankind and which have become the mode in the West. All the Roman Empire—and maybe this was one of the bases of its power——adopted children, and the adopted child had rights superior to those of the natural child; he could even become the real successor of its father. I believe that in all that concerns the regulation of birth in the broadest sense, one can find substituting satisfaction in adoption, but this requires that Western societies adopt a new attitude toward the family and that the famous voice of blood, which is more a social and cultural process than a biological one, stop being considered as a source of values for the father and the mother."

The discussion of genetic engineering and genetic counseling up to this point had revolved largely about only one side of what I saw as a two-pronged issue. The second prong certainly deserved as much, if not more, attention. I asked the assembly:

"Do we not, by opening the door to one kind of genetic engineering—intervention in disease—also open another door to improvement of the race, the door we want closed? Once the thesis is made that intervention disappears, so to speak, in the large genetic pool, the question will very quickly be raised about whether we should not also intervene if the damage is not discernible. Should we not intervene for positive purposes? As the papers we've heard today illustrate, the line between intervening to curb an illness and intervening to improve the race is not clear. I am not suggesting that we should not intervene. I am just raising the moral and social question as to where we draw the line."

"Mr. Chairman, I would like to return just for a moment to an issue that Professor Etzioni raised," Gellhorn said. "At least, as I understand his comments and his questions, it was not that he was advocating any mechanism for improving the race, but rather, he was suggesting the possibility that if we accepted the fact that there was a place for intervening or for counseling with regards to an individual, could this not then also open the door on attempts for intervention in the improvement of the race? Now, again, it seems to me that if the issue were raised before this council, about whether to establish a policy in favor of governmental regulations that would lead to improvement of the race, the answer would be very clear: no, there is no point in this; the available evidence suggests that this is absolutely without purpose. On the other hand, we do know that in our own lifetime, this was a governmental policy, and therefore it is a real question. Would it not therefore be a wise recommendation of this group to establish a kind of international commission, made up of scientists from various areas, that would be able to counsel governments, and provide them with information, so that if this issue does arise as a policy matter at a governmental level, the government in question would have the sort of information that we now have had from a group of experts? Such a group should be made up of scientific experts as well as individual sociologists in the various branches of the social sciences who would weigh the implications of any proposals, and I would consider that this might ultimately be one of the recommendations that would emerge from this round-table conference."

To my delight, Prof. Marcel Florkin, a representative of the International Union of Biochemists, joined in: "I think this is a very good proposal and ought to be taken up by this conference, possibly toward the end of its deliberations."

That was good news indeed. To have Florkin, Hamburger, and Fliedner favor a Health-Ethics Commission was very fine, but to have the stately, presiding Gellhorn endorse it was significant indeed. Could one get this international assembly to endorse it? Would it not be rather comforting—aside from being wise—to ask a council of wise people, given more time and the support of a research staff, to ponder these matters systematically rather than in a three-day, ad-hoc meeting?

CHAPTER FOUR

Should We Breed a Superior Race?

Healers versus Breeders

Often at the end of a day of scientific meetings, I feel as though I have spent the whole time standing on my toes. I am used to taking in data by reading; listening to long papers is a strain. I am used to shutting the door to my study, turning off the phone, and emerging occasionally for brief dialogues. A day of pointed exchanges in the meeting hall, during lunch and dinner—even in the rest rooms—requires a lot of fast thinking and responding on one's feet. Thus, when nine o'clock or so comes around, I usually feel rather unintellectual. I'd rather sit back and listen to a symphony or read a mystery novel than plow through another scientific paper.

Tonight was different. Under the stimulation of the preceding two days, I felt I wanted to know more and explore further. I wished I could make up, in a single night's reading, for all the biology classes I did not take in college. Above all, I found it difficult to decide what my position was on the issues underlying the many specific questions raised in the intensive meetings. Does the new technology of genetics promise an ever better quality of human beings or threaten mankind with a new source of enslavement? Would these developments be used to breed wiser, warmer people, or would they lead to a tyrannical 1984 or to a Brave New World even before the seventies were over? Would the new techniques be used only to breed out faults, especially genetic illnesses, or also to foster desired features and attributes? Are we knowledgeable enough, sufficiently wise, to make such fateful decisions?

The faceless hotel room provided no distractions. Shoes off, tie on the floor, legs up, scotch and soda in hand, I tried to figure out why I felt so overwhelmed. Was it my dealing with matters which reached far outside my fields of training? The doctors were routinely using terms such as "Klinefelter's syndrome" and "triploidy," which I had to look up in a biology textbook. My degrees in sociology, philosophy, and economics were hardly a preparation for a dialogue on medical and biological matters. No wonder several of the doctors at the conference seemed to be gradually losing their patience with me. I was a layman meddling in matters whose scientific bases I could just barely follow. I decided to be more careful, to limit my contemplations, and above all, my interventions, to social consequences (my area of specialization). For example, I would do better by not trying to form an independent judgment on whether or not the age of the father, and not just the mother, mattered in causing defective genes; and limit my concerns to the question of what social difference it would make if the father's age were a factor. I could also ask about the ethical issues that would be raised by the availability of such information. Ethics is everybody's "specialty."

Another source of difficulty for me was that the discussions at the meeting moved rapidly among several frames of reference, and located, explored, and assessed the *same* genetic interventions from different perspectives. Participants at the conference switched rapidly from exploring, say, amniocentesis followed by abortion, from the viewpoint of the parents (who may or may not wish to have a deformed child), to that of society (which may or may not be willing to put up $1.75 billion a year to take care of mongoloid children), from therapeutic goals (the prevention of the birth of a deformed child) to the use of the *same* procedures for breeding purposes (e.g., choosing the sex of the child to be born), from individual rights to society's problems, from voluntary schemes to coercive interventions (e.g., laws prohibiting the marriage of feeble-minded individuals). As a sociologist I was trained to keep these perspectives carefully apart; many of the participants at the conference moved from one to the other with too much ease.

To flag where they were, the conferees used a large variety of terms: euphenics, eugenics, eutelegenesis, negative versus positive interventions, genetic engineering, genetic surgery, genetic therapy, and so on. Many of them, I learned, partially overlapped.[1] I soon realized that it was rare to find two scholars using the same terms in the same manner. In order to figure out my position, it seemed best to start by sorting out what I was going to look at. While no fixed and fast divisions came to mind, some delineation seemed possible. After doodling on a pad for a while, a measure of organization emerged.

First it seemed helpful to separate genetic interventions used for *therapeutic purposes* (e.g., to curb sickle-cell anemia) from those used for *breeding* purposes (that is, to "order" a child with certain desired attributes [e.g., six feet tall and with red hair], the way attributes of racehorses and show dogs can be specified in advance).

Next it seemed useful to distinguish between genetic interventions introduced to *serve individuals* (e.g., parents who wish a normal child or a child of high IQ) and those used to promote *societal,* or public, policy (e.g., stamp out disease, breed wiser people).

By crossing the two dimensions the way we cross coordinates to locate places on a map, it seemed possible to locate the various issues that were raised in two days of meetings:

	Therapeutic Goals	Breeding Goals
Individual Service	1	3
Societal Service	2	4

Thus the issue of whether or not a mother should be free to abort a deformed fetus belonged in cell 1, together with other individual therapeutic questions. The issue of whether or not society should promote genetic tests and abortions to curb genetic illnesses belonged, together with other public health issues, in cell 2. The question of individuals having the right to design their next child, a quite different issue from the therapeutic one, found a place in cell 3. If society could follow a policy leading to a "better" human stock, that belonged in cell 4.

I realized later that I had to make a distinction concerning the *method* of intervention used. And so I further divided the societal section into *voluntary* controls, (e.g., the way we ask, but do not force, people to limit their family size) versus *coercive* controls (e.g., the way we make couples take a Wassermann test to rule out syphilis before they marry).

The final chart looked like this:

	Therapeutic Goals	Breeding Goals
Individual Service	1. e.g., abort deformed fetuses on demand	3. e.g., artificial insemination; parents' choice of donors features
Societal Service *Voluntary*	2. e.g., encourage people to abort a deformed fetus	4. e.g., urge people to use sperm from donors who have high IQs
Coercive	e.g., require a genetic test before marriage license is issued	e.g., prohibit feeble-minded persons from marrying

Now that I had a place for each issue, I had to figure out how I felt about each box of genetic tools.

Individual Therapy

I found the first cell generated by the intersection of these co-ordinates the easiest to deal with. In principle, like Steinberg and Fraser, I could see little reason for not providing individuals with all the genetic therapeutic services they would be willing to use. Surely, I felt, no church or government should force parents to give birth to severely deformed children, and to force into the world children doomed to a distorted, miserable life. Genetic counseling, mass screening, and amniocentesis should be available to all. Many of these genetic interventions, like many other forms of advanced medical services available today, mainly help the well-off, largely because the poor are less informed, more economically constrained, and less likely to seek medical assistance. This is one of the tragedies of our society. [2]

That not enough people are trained to provide genetic counseling is another indication of a distorted priority scale of a society yet to be fully humanized.

I realized that even these most obvious and beneficial uses—that is, for individual therapeutic goals—have some catches. Firstly, new techniques should be made available only after they are well tested. In the field of genetic intervention, there are fashions and fads and occasions when new devices are made available to people before they are effective or safe. A well-known example, was the rushing of a PKU test into use when it was still faulty, and it consequently led to rather damaging diets for previously healthy children. New genetic procedures should be examined by a powerful review board, like the kind that now reviews drugs before they are marketed. But such a Health-Ethics Commission would have to be more effective and potent than the present FDA; it would have to curb both corporations seeking to make a quick buck from new techniques, and political and medical headline seekers—the health hot-rodders—like those who rushed the PKU test through (see pages 24–25 and page 53).

Once genetic techniques were proven sound, I would curb their application only under one condition: if there were good evidence to show that providing such services to individuals would cause serious, clear, and present harm to *society*. Thus I could see limiting the service to individuals if studies showed that by providing amniocentesis on demand the genetic foundation of the human race would indeed be seriously endangered—not on some future day, when all of civilization may break down—but here and now (e.g., the way cars in some areas have increased carbon monoxide to the point where the health of the residents is damaged).

I don't recognize an unlimited priority of the individual over society, if only because the individual is part of society and needs it for his or her survival and well-being. Therefore to curb pollution in downtown Los Angeles and in some New York City tunnels, we are quite correct in limiting the use of autos and urging people to drive in auto pools or take buses or trains.

Even under this circumstance, when society's need is urgent, it is often more practical and more ethical not to make people change their preference (e.g., for using cars), but to try to use new technologies to reduce the societal cost (e.g., to render the cars less polluting). Similarly, if some new genetic interventions cause problems, we should, before we ban them, see if their side- or after-effects can be eliminated. But if such measures fail, or until they are available, society, as far as I can see it, is entitled to limit service to individuals.

On second thought, I felt that in genetics such curbing is even more difficult to justify—and hence should be more infrequently introduced than in most areas—because we are dealing, not with a convenience or even an economic need, but with a very intimate, personal part of our lives. It is one thing to forbid people to drive their automobiles downtown or to require the use of seat belts; but it is quite another to force them to be sterilized or to prevent them from getting abortions or to force women to bear mongoloid children. Thus the test for curbing genetic service ought to be even more exacting than those in other areas. Fortunately, in most genetic matters, individual desires seem to complement societal ones. Or, as the presentations of Steinberg and Fraser showed, services to individuals, such as genetic counseling and abortion, have only small effects on the societal pool of genes and, therefore, on the quality of the human stock.

After all is said and done, I would place a big plus in cell 1: new genetic therapeutic techniques, once developed, would bring much joy to parents and children, and cause little discernible harm.

Societal Therapy

But what about a situation in which the *society* has therapeutic goals but the individuals don't go along with them (cell 2 of my table)? At first glance, this may seem absurd: how can we speak about a society that has therapeutic goals *other* than those of the individual members? *Whose* goals are these? And why won't the individuals accept services aimed at improving their health or that of their unborn children?

That this is quite possible can clearly be gleaned from nongenetic areas. Take tooth decay and smoking, for example. Both profluoridation and antismoking campaigns are promoted by government. If individuals were really willing to heed sound medical advice on basic matters, these campaigns would be unnecessary. But because of irrational fears (in the case of fluoridation), or addiction (to cigarettes), society *does* enter the picture. In the case of smoking, society employs *voluntary* methods (propaganda) and economic pressure (high taxes on cigarettes). In the case of fluoridation, society uses *coercive* methods: fluoride is injected into the water mains in many communities without citizen consent or active knowledge[3] precisely because, when the issue is put to a vote, it is often vetoed.[4] Thus, society's forcing of its members to attend to their health is far from an unknown phenomenon. Could society step in, on the same grounds, concerning genetic matters?

What about genetics? Coercive genetics—the use of the society's laws, courts, jails, and policemen—to force the weeding out of undesirable genes seems to me intolerable and repugnant. (Not everybody sees it this way. Dr. Harvey Bender, of the Biology Department at Notre Dame, said, in reference to voluntary sterilization, "I'm apprehensive about either form of sterilization—but how else could society enforce its genetic standards?")[5] Unlike excessive use of autos, or even the abuse of one's teeth, the union between two persons which gives life to a third one should be kept free of all government intervention. I would be horrified if the government budgeted the number of children per family or put a contraceptive drug into the water supply or forced mothers to abort "surplus children." One need only think of what would happen if some official decided that in order to reduce criminality, chromosome tests on all pregnant women would be required and abortions demanded of all mothers who carry XYY "criminal" fetuses. We would end up with policemen dragging women to abortion clinics and mothers going underground to protect their embryos. If the government uses its force with respect to these matters, it would constitute the ultimate violation of the contract which keeps people tolerant of the state. It would completely undermine the legitimacy and the moral basis of government.

Because genetic technology will improve, the appetite to interfere may well be excited. Therefore if any attempt is made to move in the direction of coercive intervention, I favor it being met with the utmost opposition by citizens and their representatives. To symbolize and ingrain the rejection of forced genetics, I would welcome the repeal of all genetics-by-legislation, that is, by force of law, which now exist. These include laws that forbid marriage among the feeble-minded. In the early part of this century, many states prohibited such marriages; today only the states of Washington and North Dakota prohibit them between men of any age and women under forty-five, if they have a history of insanity, are feeble-minded, or are imbeciles, habitual criminals, or common drunkards.[6] Denmark, hardly a socially backward country, requires sterilization of women whose IQ is less than 75.[7] Seventy-one thousand mentally retarded persons, it is reported, were sterilized in one state alone—North Carolina.[8] (That such laws still bite can be gleaned from a recent U.S. court decision that ruled to take away and hand out for adoption the twin children, aged four and one-half, of David and Diane McDonald, because in the court's view, the parents' low IQs—74 and 47, respectively—did not enable them to take proper care of the children.)[9]

The way society actually feels about these matters is well reflected in the fact that these laws are almost never enforced. Having them on the books did not matter much when they did not set a precedent for other types of interventions, because there was not much that citizens could be coerced into doing in these matters. But as the possibilities of coercive interventions are now rapidly multiplying, wisdom calls for setting up the strongest possible barriers against them.

I oppose forced genetic interventions, not the setting and promotion of genetic goals. Individuals are often shortsighted or selfish; they act as though they are the only ones in need and disregard the fact that what might work for one often will not work if all individuals act in the same manner. Hence there is good reason to take into account societal needs—those future, aggregate, and shared needs of the people who make up a given society. But these needs should be met not through the use of force, but through voluntary means. Which of the variety of voluntary means should be relied upon depends on the circumstances. If the societal need is very urgent, if the burden is overwhelming, I can see the use of economic means. For example, a government under great economic pressure, say, in a severe recession, where there is great demand for societal services, or during a prolonged health crisis, might inform its citizens that the public institutions will no longer allow parents to dump deformed children on them. Thus, while no one should force parents to abort a mongoloid fetus—now that parents have a choice—society does not have to, under all economic conditions, pick up the tab for the upbringing of such children.

When societal needs are less pressing, persuasion—without economic sanction—should be chiefly employed, such as the kind used to get people to accept birth control and to curb drinking. I can imagine ads saying, Give Your Child a Chance to Live a Full Life—Check Your Genes, or You No Longer Must Bring a Mongoloid into the World. Other educational means, even organized tours for prospective parents through wards of deformed children, could be employed. Mario Biaggi stated that in his thirty years of public service, first as a New York policeman and later as a Congressman, nothing moved him more than the sight of retarded children in public institutions. He reports about one of them:

> Adrienne, I found, was a mass of injuries—unhealed sores and bruises over her entire torso, one eye swollen and closed, chin stitched, nose badly scraped, possible skull damage. She had been abused, allegedly by other patients. [10]

Public health officers may not wish to rush out wearing slogans—Let's Stamp Out All Genetic Illnesses the Way We Abolished Polio, or even Take Your Test and Be Free of Mongolism—because genetic illnesses cannot be overcome the way other diseases can be. Testing everyone's chromosomes and pulling out all the sick genes will not rid us of genetic illnesses. The basic reason is that nature continues to produce new supplies of such genes through mutations. To my mind, these are like printing errors; however carefully you set up the type, even if it is free of errors, for every x-thousand print-offs (or children to be conceived), a certain number will be defective. Thus even those genetic illnesses that result in the early deaths of all carriers *before* they have children do not disappear.

It has been suggested that the whole attempt at genetic public policy is a waste of time, that it is hopeless. [11] However, this is clearly not the case. While we may not be able to reduce the defective rate, we can at least catch nature's errors and eliminate them before they turn into miserable children, agonizing parents, and public charges. We may well have to repeat the process for each generation, but this does *not* make it without value. While it would be preferable to eliminate these illnesses ónce and for all, the next best thing is to eliminate their consequences—their human and economic costs.

As I see it, then, the basic rights of an individual in a free society include that of having as many of whatever kind of children a person is willing to have; society can try to persuade people to have fewer children or to abort severely deformed ones, but it cannot force these choices. However, the individual's rights do not include the liberty to charge the upbringing of their children to the public. I can picture a society going so far as to inform all prospective mothers, especially those in high risk categories, that a genetic test is highly advisable, and further, to inform those whose tests show them to be carrying a deformed fetus, that *they* will have to provide for it. But, to repeat, the use of genetic police or inspectors must never be tolerated.

With the help of my chart (page 104), I sorted out my feelings about genetic interventions for therapeutic purposes, and summarized them. On individual demand (cell 1)—yes, by all means, as long as the procedures are medically sound. To advance public policy (cell 2)—yes, by voluntary means; but not by coercive means under any circumstances. This leaves the other half of the exercise, the use of genetic interventions, not to remove or otherwise overcome sickness-producing genes, but to promote those genes which are believed to carry desired qualities, ranging from a more attractive complexion to higher intelligence.

Societal Breeding

The use of genetic techniques for improving genetic qualities raises a quite different set of questions. This form of genetic intervention is discussed most often from a societal viewpoint (cell 4) because it is here that the best-known attempts at breeding "better" people have been made. In the past these efforts were directed toward goals that almost all people find abhorrent, especially those by the Nazis. They tried to breed a "master" race, using such abusive techniques as the extermination of those whom they felt had inferior genetic qualities (not just Jews, but also feeble-minded Aryans and other populations). Within the German population, the regime imposed the compulsory sterilization of manic depressives, severe alcoholics, the feeble-minded, epileptics, and those suffering from hereditary blindness and deafness; and the castration of dangerous and habitual criminals. To preserve the hereditary soundness of the German race, marriage was forbidden when one of the parties had a dangerous contagious disease or suffered from mental derangement or an hereditary disease. Racial intermixing and intermarriage of Germans with foreigners were prohibited, and a German who did so could lose his status as a German.[12]

The taboo we now have against the deliberate breeding of certain types of people is so effective that just thinking about it made me feel ashamed and somewhat defensive. The label "racist" came to my mind. It is used so often to refer to people opposed, openly or covertly, to equal rights for minority members, especially to those who base their position on inherent, genetic differences between the majority and the minorities. Thus, aside from reminding one of the Nazis, the very notion of selective breeding brought to mind the Ku Klux Klan.

But the meetings had their effect; having been trained as a scientist *not* to take prevalent societal "no-no's" at their face value, I was also curious. Was it time to examine these taboos? Could one simply dismiss out of hand all the "promised lands" that distinguished scholars such as Bentley Glass and H. J. Muller have pointed out as within our reach? Mankind was in sore enough condition. It seemed rash simply to brand "unthinkable" the promise of breeding a race who would have "freedom from gross physical and mental defects, sound health, high intelligence, general adaptability, integrity of character and nobility of spirit," as Glass put it.[13] Why dismiss the notion of using biotechnology to create people with "a genuine warmth of fellow feeling and a cooperative disposition, a depth and breadth of intellectual capacity, moral courage and integrity, an appreciation of nature and art, and an aptness of expression and communication," [14] as Muller has it. It seemed presumptuous to ignore these statements, even though I did feel kind of wicked even thinking about them.

Also, I felt that I could not easily dismiss the argument that the pendulum of public policy had swung much too far in the educationalist and revisionist direction, away from biological considerations. In typical dialectic fashion, we had moved from the thesis, popular in the first decade of this century, that man is governed by biological instincts (sex, hunger, or aggression) to the antithesis of the educationalist concept of man. In 1925 John B. Watson, a founder of behaviorism, issued his famous challenge:

> Give me a dozen healthy infants, well-formed, and my own specified world to bring them up in, and I'll guarantee to take any one at random and train him to become any type of specialist I might select—doctor, lawyer, artist, merchant-chief, and yes, even beggarman and thief, regardless of his talents, penchants, tendencies, abilities, vocations, and race of his ancestors. [15]

Other educationalists followed, especially once the instinct theories became associated with fascism in Europe and racism in the United States. By the late Sixties, numerous programs, from labor training to compensatory education of the disadvantaged, from mental health clinics to cures for smokers, all assumed that education could readily improve the lot of anybody. [16]

When it became increasingly evident that this assumption was not always valid, the interest in physiological and genetic factors was reawakened. The Coleman Report, published in 1966, probably marked the turning point. It raised difficult questions about the potency of education. Then came Arthur Jensen's and Richard Herrnstein's articles, which argued that IQ differences between blacks and whites were, to a significant extent, genetically inherited. Each of these documents had a much larger impact than a typical scholarly work, because they were publicized, discussed, and debated, and because the era was ripe for a reaction to education.

The new era is unlikely to return us to Fascist notions of genetic determinism, but instead, will move public policy toward a synthesis which would rely on both educational and biological factors. The synthesis era will be concerned with their combinations and interaction. Thus, compared to the educationalist period which is ending, the new era seems to show more interest in and tolerance for genetic engineering than ever before, but without going overboard and seeing it as a cure-all.

Finally, one cannot dismiss out of hand the notion that our drive to govern our condition, rather than being subject to the blind fluctuations of forces we can neither understand nor control, might be helped through biological engineering in addition to institutional reforms and power redistribution. The curse of modernity is that the revolutionary expansion of means—of instruments—have rebelled against the creator and his purposes. Like a Frankenstein's monster, technology has gone beyond the control of its maker; it distorts society to fit the logic of instruments rather than to serve the genuine needs of its members. The primary mission for the next era is the restoration of the priority of human values. This may be reflected, for instance, in the willingness to trade off at least some economic growth and technical progress for more humane work and a greater care of nature—in short, there may be a less competitive society.

The trouble is that, at present, all efforts to restore the primacy of human values over tools—by expanding our brain power and wisdom—have not progressed very far. Efforts to do so via institutional reforms, social revolutions, or rejuvenation of self seem to provide at best only partial solutions. Hence one has to consider the notion, advanced by Glass, Muller, and others, that a "higher," less aggressive, more intelligent breed may have to be biologically cultivated before a more humane society can arise. As one biologist put it:

> From the point of view of genetics, man is a barbarian clad in the trappings of a civilization in which he is ill at ease, and barely able to contend. Social scientists pin their hopes for easing this unfortunate state on the possibility of improving human institutes [he probably meant institutions] and environments. But with human genetics as it is, this is a most dubious proposition. [17]

The author goes on to define the sources of our problems as nationalism, aggressiveness, and excessive bureaucratic inclinations, all of which render society unmanageable. He suggests that genetic engineering could help remedy all these proclivities.

It is easy to ignore both the snide remark about social scientists and the overenthusiasm of the position, and focus on the author's basic thrust. It is downright naive to believe that in large organizations, such as federal bureaucracies, the tendency to malfunction has a genetic base. But other attributes, such as aggressiveness, intelligence, level of energy—and hence achievement motivation—aren't these *in part* genetically affected? True, a person who is aggressive could be trained, that is, educated, to be a prosecuting attorney, a soldier, or an assassin, a point stressed by the educationalists. But isn't it also true that, given an aggressive race, peace is going to be difficult to achieve and sustain whatever the educational reform efforts? And if people tend to be lethargic, can education turn them into a productive and creative race? It seems to me that we must draw on both societal and biological factors if the human condition is to be bettered. This conclusion may be obvious to some, but it is hardly so to those brought up in the mainstream of the libertarian or social-science traditions of the last generation.

Is such thinking "racist"? Yes, if one assumes that some groups of people have "bad" genes while others have superior ones. But racism isn't the issue if, as I see it, *all* of mankind's genetic stock may well stand in need of improvement: no one group or race has a monopoly on good or bad genes. Human breeders will be like those who, seeking a superior breed of cattle, try to combine the superior qualities of several existing breeds; the resulting hybrid will have little resemblance to any of the original races.

Secondly, the human hybrids who result will not necessarily be popular, and hence, those not so endowed will not be stigmatized. As one author describes the "superman" to be bred, he will have the nose of a bloodhound, the wings of a dragonfly, the gills of a fish, the ears of a snake, and will be clothed in seasonal body hairs, with the metabolism of a tape worm.[18] Each of these features is quite useful; for instance, the metabolism minimizes waste and the need for elimination. But I don't believe I am being uncharitable if I say I do not expect mothers to line up to breed such superbabies.

Thirdly, just as no group has all the desired genes, so each subpopulation seems by nature to be afflicted with one or more undesirable genes. Blacks are afflicted with sickle-cell anemia; Eastern and Central European Jews have Tay-Sachs disease; Mediterranean stock, Cooley disease; Caucasians, cystic fibrosis; and so on. And even if there is a group that is worse off, this calls for more compensation and public service, not racial slurs.

But I could hear Moltmann arguing, like others before him, that such a breeding policy was possible for racehorses, hogs, and dogs, but not for human beings. We could never agree on what to breed: athletes, eggheads, redheads or blonds. The very attempt to do so would break society up in conflict.

I myself had believed, when I came to the Paris conference, that a breeding policy would impose just such a strain on society. But now, on second thought, I asked myself, if we could help grow, say, more intelligent and warmer persons, wouldn't most of us want to do so? And who said that we need a uniform race? Could we not breed some of each kind? Above all, since the implementation is to be *voluntary,* there will be no more uniformity than people will choose to have. And no need to reach a consensus.

At issue is a public policy which welcomes certain biological features over others—e.g., energetic over lethargic qualities. This is rather similar in nature to our call for limiting the family size. Some are influenced by it, others ignore it. Similarly, in the case of breeding policies, even if there were one recommended fruit, many would not buy it. And if some attributes do prevail—as we do *now* breed taller people because of wider acceptance of the recommended use of vitamins—it would be only because many people accept the policy. In short, I no longer saw a contradiction between a genetic policy and a democratic society.

I was aware that several biologists have argued that this is all pie in the sky, that such breeding is technically infeasible in the near or even remote future. The whole argument is unnecessary, they say, because no such changes can be effected. First of all, as mandatory and uniform policies are out of the question, voluntary adherence would be limited in scope and hence in effect; second, to have the desired effect, some rather specific strictures would have to be adhered to. The example used concerns intelligence (an unfortunate choice, because intelligence is affected by several genes, and the effect is surely more complex and difficult to bring about than, say, changes in height).

These skeptics point out that in order to achieve the desired effect, women with low IQ's would have to marry men with high IQ's, or men with low IQ's women with high IQ's (the mating of two high IQ's does not make for a higher IQ), which is unlikely to be carried out voluntarily on a large scale; and—to repeat—coercion on behalf of such a goal is unthinkable and repellent. As to the use of artificial insemination, if 10 percent of the women of one generation used the sperm of 160 high IQ donors, the average IQ rise would still be a very low 1.5 points.[19]

While I did not feel competent to judge these matters, I couldn't help but note that other equally distinguished scientists, such as Glass and Muller, were much more optimistic as to what could be done through genetic engineering. It has also been pointed out that while the average increase in IQ that might result from genetic engineering would be low, a mere increase of even 1 percent in the average would result in 3.5 to 4 million additional very high IQ (175) persons.[20] Maybe your next son or daughter would be one.

Though I could not document my feelings, I felt that the truth probably lay, as it so often does, somewhere between the extremes, between the advocates and the deriders. I concluded that unless someone could bring up a new argument against a public policy that would encourage people voluntarily to favor certain traits—say intelligence or warmth—and if genetic promotion of these traits was technically feasible, I wouldn't oppose it. I surely would not like to see a federal crash project invest five billion dollars to breed a brighter or more peaceful race, but a limited genetic experiment might be acceptable. It has been argued that even such a limited experiment would require twenty generations to complete because of the slow accumulative effect of such changes. The scope of the effect depends, of course, on how many people will choose to heed the genetic suggestions and the extent to which they will marry each other rather than "outsiders." If many participate and marry among themselves, the effect will be greater. And while the benefits will almost surely be gradual and not sensational, this is hardly a reason to oppose them.

I was jotting these thoughts down on a pad late into the night. My glass was dry, so I made my way to the bar in the back of the hotel lobby. During most international meetings I've attended, more issues get argued out in hotel bars than in the conference rooms, and you can find one participant or another at the bar practically any time—usually not smashed, but slowly sipping his drink between wordy exchanges. This time I found Dr. Case, sitting by himself in front of a large beer. He waved me over. I asked him if I could try out on him a thought with which I had been struggling. He promised me his undivided attention and I explained why I had begun feeling that one couldn't easily dismiss purposeful efforts to improve the race.

"Oh, my," was his first reaction. He was silent for quite a while and then added: "Would you like to live in a state like that?"

At first, I didn't get his point. "Tell me more," I asked. "Why not?"

"How would you be sure the government would promote the genes you favored?"

As with many others, the notion of breeding brought to his mind the horror of coercive measures rather than the voluntary steps we use so often in other highly personal matters.

"I guess, on this one, I agree with Lederberg," I said. "He says that such a policy would be *preceded* by tyranny. [21] There are 'coercive eugenics,' those imposed by the state, which must be abhorrent to all, and must be fought by all means known to human beings, like other totalitarian policies. And there are 'voluntary eugenics,' publicly promoted, freely accepted or refused, like participation in the March of Dimes. One ought not to confuse the two."

We sat silently for a while. Case looked at his watch and declared that it was high time to retire. I stayed up, trying to complete my exercise. I had covered all the parts of my scheme except the one least often explored: what is one to say about individuals seeking to breed children the way gardeners seek an attractive hybrid of flowers? One may be reluctant to favor a public policy on the side of genetic improvements—but how about individuals shaping the genes of their offspring to their own heart's desire?

Individual Breeding

"Gene shopping," that is, choosing and combining the biological qualities of a child yet to be conceived and designing it to the parents' preferences, is discussed chiefly in science fiction and occasionally in the popular press. Because the technical means for gene shopping are not available now and will not be in the near future, scientists rarely view these questions seriously. To most experts in the field, the day when a parent can go to a gene-mart and tell the clerk to give him the genes needed for a blond, blue-eyed, tall, slender, high-IQ boy is so remote that they feel that it's quite unnecessary to worry about the wisdom, ethicality, or social consequences of developing genetic supermarkets.

Experts maintain that gene shopping is a long way off because most of those biological attributes which are genetically determined are controlled not by one, but by several genes, acting together in ways far from fully understood. Thus it might be relatively easy to shop one day for height or hair color (attributes relatively simply determined), but it is quite unclear what would have to be bought if one were seeking a high IQ or many other desirable attributes.

Secondly, most attributes are affected by both the genes in the husband's sperm *and* in the wife's eggs. Hence, unless the trait is dominant, you cannot shop for it without being willing to gamble. Some writers talk about buying the egg, too, and implanting it after it has been fertilized with the desired sperm, or about frozen embryos made to specification. But technically, such possibilities are even more remote than gene-shopping.[22] Moreover, since these techniques require not just the use of easy-to-get male genes, but also surgical extraction of eggs and surgical replantation and fertilization in a laboratory, the procedure raises many more ethical issues than gene-shopping.

Thirdly, most of our attributes are shaped in an interaction between our genetic inheritance and our psychic and social upbringing. The shoppers who ask for high-IQ genes may be quite disappointed when they find they have a clever but unmotivated, or smart-alecky, child, or one who misapplies his or her talents, or looks down on his "dumb" parents.

Last, but not least, we know from breeding domesticated animals that if we push one attribute, we tend to weaken most others, ending up with a highly vulnerable, unbalanced species. Thus, poodles bred to promote their woolly coats, tend to suffer from severe ear troubles.[23] (Pushing a particular trait is achieved through pure breeding of a type, which is done through inbreeding; inbreeding inadvertently intensifies other genetic traits which may be genetic deficiencies.)

All these technical difficulties will have to be faced by all breeders, whether they breed to advance a public policy or to cultivate parental desires. However, they will particularly limit what an individual can achieve. A society can benefit from aggregate, "statistical" benefits; for example, if its efforts lead to an average improvement, even if it is not manifest in each person, and even if it is highly diluted by other, nonbiological factors, it still may be beneficial. But to parents interested in specific attributes for their next child, statistical changes over a whole generation of children are of little interest; they want *their* child to be brighter, taller, or whatever, and individual change is particularly difficult to achieve. All in all, then, gene shopping seems to hold much less promise than the press or its outspoken advocates have claimed.

But it does not follow that one need not discuss the implications of breeding to individual order, because *some* gene shopping is technically possible right now. While the available procedures are very primitive and costly, the issues they raise, at least psychologically, are not different in principle from those raised by "future" developments. Indeed, precisely because of early technical and related social developments (social acceptance or disapproval often affects later innovations), the question must be faced, and the sooner, the better.

Sex-choice is a case at hand. The *same procedure now used to control mongolism*—that is, the combination of amniocentesis with abortion—can today secure an infant of the desired sex. When a test is made to determine if the fetus is mongoloid, its sex is often determined as well. Many doctors will not inform the parents of this finding because they do not wish the sex of the fetus to be used in abortion consideration. [24] There seems to be, so far, no case on record in which a doctor agreed to abort a fetus because it was of an undesired sex. However, one doctor reported that he was tricked into doing so by parents who asked for amniocentesis to check against mongolism; told they had a normal female fetus, the parents proceeded to arrange for an abortion, for they wanted a boy. [25] But just as a doctor can be found for any other illicit purpose, it is just a question of time before this is done—if it has not been done yet, off the record. Also, the development of sex choice techniques may make it possible in the *near future* to choose sex without abortion, by separating the male-producing sperm from female ones [26] (which would then require artificial insemination), or by providing the woman with a douche which would be inhospitable to one of the two kinds of sperms, (the female providing X or the male making Y), which would make sex-choice as easy as taking a medication. [27] Success is more likely to come about here than with most traits, because the genes in the mother's egg play no active role in determining the outcome. Thus sex choosers have to deal with only one ingredient rather than two and with the more manipulable of the two.

The question, then, cannot be avoided: Should parents be allowed to choose their child's sex, and, by implication, other genetic qualities, a choice which obviously is not one of health over illness, not a therapeutic matter at all, but clearly one of breeding—should they be allowed this choice the way a person may choose a Doberman over a Poodle or the other way around?

As long as sex choice entails abortion, one may say that, because of the marginal risk to the mother (and her future children) that is involved,[28] this procedure is tolerable for therapeutic, but not for breeding, purposes. The same holds for the question, who should have the right to decide if the risk is acceptable?

We already allow—indeed, at least indirectly encourage through our stop-the-population-growth propaganda—parents to abort children in order to limit their family size, hardly a therapeutic goal. We say that both parents and children would be happier if there are no more than two children per family. Now, should doctors or the state decide that parents are not allowed to plan their children—their sex and, soon, other attributes—only their number? And what if a family of four boys feels one girl is essential to make it happy? It seems to me the decision should be up to the parents.

When I came to the Paris conference I felt that the aggregate consequences for society of its members freely choosing the sex of their children would be quite undesirable (see pages 16, 29–30, 48 and Appendix 7). This still seemed valid to me. But now, on second thought, these consequences seemed no longer severe enough to warrant limiting the development of sex-choice techniques, even if we could curb them or prohibit their application for breeding purposes. I recognize that society has needs of its own and that a severe sex imbalance could damage these; by my calculations, however, the damage would not be considerable. I had calculated an imbalance of 7 percent-per-year male surplus. This has almost surely declined with the impact of Women's Liberation, which has come about since those calculations were made.[29]

Thus I hold that since society is not likely to be seriously undermined by such techniques, we should not prevent individuals from gaining whatever happiness they can. If this entails adding a boy or a girl to their family, why not let them?

It seems to me the same holds true for other means of genetic shopping now available. Artificial insemination (AID) is used now to help infertile fathers. No information about the attributes of the donor is usually given to the parents, although doctors tend to choose one whose features are similar to those of the parents. For instance, they do not give a white mother the sperm of a black father or give to a black mother the sperm of a white father.[30] However, most doctors would refuse to provide AID to fertile fathers or allow the prospective parents to choose among donors according to some desired attributes, say, asking the doctor to find a tall donor if the parents want a tall child.

Here the risk is not medical but psychic. In some cases fathers later resent children born as a result of artificial insemination, and mothers have been known to develop romantic infatuations for the unknown biological father. These cases are used by some to support the thesis that artificial insemination should be used only when all else has failed, and not for the biological designing of children. But, firstly, data show that these tensions can be handled by carefully explaining the issues to the couples involved and by providing psychological counseling. In any event, the cases of serious emotional trouble are very rare. [31] Secondly, the individuals' happiness and their preferences should prevail. Doctors may well alert parents to the psychic dangers (as they do when the father is infertile), but beyond that, it should be up to the prospective parents to decide. (Doctors who feel AID violates their personal ethics should, of course, not be forced to provide it, any more than a Catholic doctor should be made to provide an abortion. However, if the health authorities and medical societies make the service legitimate, prospective parents will easily find many quite willing doctors.) Thus, if a couple who are short in stature feel very strongly that they don't want to impose such a condition on their child, why not allow them to get sperm from a tall, anonymous donor?

Parents may initially expect too much. Many attributes are not inherited or they are inherited only in part, and a sperm contains a large variety of genes. A mother with an IQ of 100, getting the sperm of a 140-IQ donor, cannot be assured that her child will have an IQ of 140, or 120, or even higher than 100. Moreover, as long as gene-by-gene shopping is not possible, the mother must "buy" sperm—the whole package of genes, which may include attributes she does not seek. To offer parents the "packages" they desire, sizeable sperm banks would have to be established. Nevertheless, some attributes can be ordered, and for others, one can take a gamble, which is what we do when we use the natural way. If we allow people to gamble on winning a fortune in a state lottery, it seems to me that we couldn't prohibit them from trying to improve the biological lot of their children.

The next step may well be that sperm banks, which already exist—they have been set up for fathers who wish to preserve some of their sperm when they undergo vasectomies or are exposed to radiation in their work [32] — will be used to store sperm from anonymous donors, typed according to their attributes. Parents or unmarried women would prepare a list of specifications for their next child and give this list to a sperm teller at the sperm bank who would check the files to see if all the specifications or only certain combinations of specifications were available. A fee would be paid and a vial would be issued. Sperm might one day be available in the form of a suppository that could be used without a doctor's assistance. Individual breeding would then be on its way and might even become fashionable.

Two matters should concern us if we proceed in this direction. The first is that prospective users be well informed as to how much can be achieved through AID. If sperm shopping catches on, it surely will be necessary to educate the public on what to expect, a kind of consumer education to discourage excessive expectations and tensions which might result. And the various consumer watchdogs should make sure that the sperm banks don't engage in false advertising, oversell their product, or mislead prospective parents.

Second, one must consider the notion that these genetic interventions, which rely on artificial insemination, are a form of adultery, an immoral union. Theologians suggest that AID, especially when the father is fertile, will further undermine the family. Many religions see it this way, and so do large segments of the public. A 1969 national poll shows that 49 percent of men and 62 percent of women accept the idea of artificial insemination but only in cases where the husband's sperm is too weak to work without a doctor's assistance. [33] The use of an anonymous donor was approved by 19 percent; 35

percent approved if this is the only way a family can conceive or have normal children.[34]   But adultery without AID is very common anyhow. (The same national poll which gauged public attitude toward AID also found that 50 percent of the respondents said that they knew a husband or a wife who was unfaithful.)[35]   Society does not see fit to enforce the law forbidding adultery. *Motels* would have to be banned long before sperm banks, if society ever wished to really take on adultery. Also, when both parents consent, and the procedure is conducted in the cold medical sterility of a doctor's office, such a union has little to do with assignations. In short, while I would not promote breeding for personal purpose (e.g., get yourself a blond child), and I do see a need to protect individuals from sperm salesmen, I see no reason to outlaw or stigmatize such a procedure. If it had been available today, my wife and I might well have used it to add a daughter to our present family of boys.

I continued to write down my thoughts on a pad, as I sat in the corner of the bar. By now, Dr. Case was probably sound asleep; the hotel was very quiet. There were only two other persons besides me still sitting at the bar. The waiter plopped the bill in front of me, implying either that it was time to close or that I was not consuming enough. As I left, reviewing the evening's notes, I was rather satisfied with my exercise. I had been able to sort out what I should favor, what I could tolerate, and what I had to oppose: Genetic interventions for individual therapy needed more support, not less. Societal force should not be applied for either health or breeding purposes. Voluntary promotion of public policy in genetic matters, for either of the twin goals, made good sense, though not much could be gained for society because of technical limitations. Individuals should be free to breed what they wanted to. Steps, however, must be taken to see that the public is better informed as to what to expect, so that they can make wiser decisions.

I could almost hear some of my more radical colleagues saying: "One must question this primacy of the individual; it all sounds preciously like advocating strip-mining." However, strip-mining is not carried out by "individuals" pursuing their personal needs, but by profit-seeking *corporations*. And if corporations get into the genetic picture, I surely do hold that they need to be regulated.

Like advancing troops skirting a well-entrenched enemy position,

I had left behind a tough question captured in the slogan: One Thing Always Leads to Another. If we open the door to genetic engineering of one kind, will it not lead to other, less desirable forms? Would not voluntary forms turn into coercive ones as governments become attracted to prospective gains?

The Slippery Slope, or The Postvirginity Problem

Neat conceptual charts, I was quite aware, do not keep social forces within their boundaries. Actually, one of the arguments most often made against genetic engineering is that to engage in it is to step on a slippery slope; once you lose your footing, you find yourself on your backside at the bottom of the slope. Dr. Watson suggested that once the taboo against experiments with fetuses were violated through continuation of the *in vitro* experiments, "All hell would break loose." [36] Some fear that once we use genetic interventions for therapeutic purposes, they will also be used to breed people. Others think that once they are allowed for individuals, they will also be used by governments. Still others feel that what might start out as a voluntary public policy in a free society may end up as a coercive measure used by a devilish totalitarian government to xerox a million copies of supersoldiers or secret agents. Finally, a slippage of morals is feared. Once we start making babies on assembly lines, what will happen to the family—indeed, to the sanctity of human life?

Little hope for avoiding the problem could be expected from the techniques themselves, because the interventions which work for one goal could also be used for the others. Like master keys, they can serve to open a large variety of genetic doors.

Thus genetic techniques—counseling, amniocentesis, karyotyping, sending messages to genes via viruses, and so forth—could serve as both healers and breeders, to advance individual needs as well as societal goals. Cloning is a possible exception; I see no therapeutic value in it, though its advocates say that it could save lives by growing spare parts for a person. Thus, if your kidney malfunctioned, a replacement could be taken from an identical body; and such transplantation of genetically identical tissue is much safer. One need not grow a clone to order and kill it off to take the needed part; rather, from a group of a thousand identical persons, parts from those dying of natural causes could be used much more effectively than transplants from non-identical donors. It would be as if we all had many identical twins.

But what is a society to do? Not step on the "slope" at all? No one should oppose all genetic interventions, as the Church does, because this stand wreaks misery on people, misery we now know how to avert. The act of not enabling parents to choose in order to avoid bringing up a mongoloid child has drastically changed over the recent years, not just medically but morally. In the past, doctors knew only that a child of certain parents would have a certain chance of being defective. Thus if the mother was between the ages of forty and forty-four, and an American, her chances of having a mongoloid child were one out of eighty.[37] This gave the parents scarcely more specific information about a pregnancy than they'd had at first, and left them with only harsh options.

To illustrate: imagine if, at that time, you had been considering having a child. After having been told that it had one out of eighty chances of being severely deformed, you were further advised that abortion could not help, because your next pregnancy would have the same (actually slightly higher because you'd be older) chances of being affected. Your choices, then, if you wanted to have a child of your "own flesh and blood," would have been limited to either gambling on its being normal or not having one at all.

Since the early Sixties, however, with the advent of amniocentesis, doctors can tell you with very high degree of accuracy whether a *particular* fetus is apparently normal or grossly defective. Parents now can literally choose between a mongoloid child and a normal one, and know that if they abort an afflicted fetus, they can very likely replace it with a normal child! It is like the difference between having to go for the jackpot each time, with life at hard labor if you fail, as compared to putting your money on a horse in a race in which there are only two horses in the running, after having been told, with 99 percent accuracy, which is to be the winner! It seems inconceivable that such information and medical help would be withheld in the name of such an abstract notion as a slippery slope. Thus, morally, I believe, we have no choice but to set our foot on the slope and negotiate it part of the way. We shall have to find ways to stop from rolling down, but we cannot afford to allow fear of the lower reaches to keep us from reaping the benefits awaiting us at higher reaches.

Fortunately, the various usages to which the same genetic technologies are put are quite clearly linked to separate goals. This eases the problem of slippage. For instance, doctors are applying amniocentesis-plus-abortion for one purpose (therapeutic), but not for the other (breeding). While I would allow its use for both purposes, the point is that if one wants to draw a line to indicate the acceptance of one and the rejection of the other, it is obviously quite possible to do so.

The other slippage about which various commentators worry lies in the means used to promote public policy in this field. A government, it is said, could start with completely voluntary efforts (the way we now try to persuade people to have smaller families), go on to exert economic pressure (say, in the form of tax penalties for those who do not comply), and from there take outright coercive steps (mandatory sterilization). There is no denying that there is such a danger. Governments do have a tendency to do foolish things and to rush in where angels fear to tread. But one must also note that such slippage is not as inevitable as those who point to it often imply. There is public debate, a political struggle, and even moral assessment, whenever means of control are changed—for instance, when America legalized abortions (a change from coercive to voluntary control) or moved to ban smoking ads (from voluntary to coercive), etc., etc. To draw an example from another area, one need not oppose equipping cars with fume controls as a mandatory depollution measure out of fear that it will lead to the banning of the use of cars, although the first step could be followed by the latter.

Finally, some observers worry about moral slippage—a gradual expansion of the purposes for which people would find genetic interferences acceptable, as distinct from expanding government encroachment. There are now several mores agitating against the notion of genetic engineering, including those which extol the family, the taboo against recognizing racial differences, and the values which stress the sanctity of human beings. Once tampering with all these becomes acceptable—through *in vitro* breeding, etc.—there would be a further weakening of the moral veneer which keeps people civilized and keeps them viewing their spouse, fellow citizens, all persons, as ends rather than means. These might be affected by genetic engineering, although there are scores of more powerful factors that challenge them.

However, the historical record shows that while some taboos surely did gradually weaken (most religious do's and don't's in the modern era), others were reset—that is, removed from one point (e.g., virginity), and set up at another (e.g., in favor of one liaison at a time and against one-night stands). Such revised taboos, it may be said, don't bind well; there are quite a few who violate them. But the same holds true for the old ones.

In short, slippage is not a foregone conclusion. Our choice is not limited to a technological free-for-all or a conservative clinging to traditional norms.

We have more options than to remain fearfully stuck at the top of the slope, afraid to pick up fruits growing midway, out of fear of ending up at the bottom on our collective rear end. If wiser public policies are to be formed, we must further evolve our capacity to reset taboos instead of being frozen in by them or breaking down into anarchy.

To close this matter, I felt two subsidiary questions remained to be answered. One concerned the role of technology in bringing about the expected social changes: would they follow willy-nilly, once genetic breakthroughs took place, or did we have a choice? The other question concerned the ways and means of resetting taboos.

The Technical Factor in the Social Mix

It is said we have no choice but to proceed, since the knowledge is already available. We've already lost our innocence. What John F. Kennedy once said about nuclear physics, we must now say about the new genetics—there is no way to return the genie of knowledge into the bottle. Thus no one can make us forget how to conduct amniocentesis. We may try to regulate its use, but we cannot make it disappear. Nor could anyone stop future basic research on genes and ways to affect them, any more than the Church could have stopped Galileo. And the commercial interests in the breeding of animals are so powerful that practical applications of this knowledge to mammals is sure to continue. Finally, some applications to human beings are already in hand. Our choices are therefore limited to reviewing, editing, and encouraging some uses; discouraging others; and opposing still others; but we cannot taboo the whole field.

Several distinguished geneticists argue that the growing concern about the social and moral implications of genetic engineering is highly premature. Especially adamant on this point is Joshua Lederberg, whose words have stayed with me: "Should we spend much time worrying about the ethical implications of the genetic findings of the next century, when we must do this on the basis of a set of assumptions about the human condition that will surely change dramatically in every other way?"[38]

Lederberg points out that we do *not* know how to do most of the things whose merits are so hotly debated, from the creation of artificial wombs to asexual reproduction. He holds that "according to journalistic accounts, we will shortly be writing prescriptions for human quality to order. 'Do you want your baby to be eight feet tall or have four hands? Just tell the geneticist, and he will arrange it for you,' goes this line of advertisement. But the most sophisticated geneticist today is baffled by such challenges as Huntington's disease." [39]

Lederberg seemed to me quite right, up to a point. Surely both those who try now to plan for the year 2000 and those who oppose steps to be taken now, on the basis of assumptions about what the twenty-first century will be like, are being unrealistic. The record of such projections is so poor that, though they may serve a variety of nonscientific purposes—from titillating journalism to clarification of our *present* values—they should not be used to guide public policy now. On the other hand, though it is rather foolish to oppose helping infertile women because one day people may be xeroxed, there is a sufficient number of genetic interventions taking place today to give one pause concerning *their* implications. Attempts to grow babies in test tubes are being carried out now, and no one can be sure that the prospects of a baby born to a machine, or a fetus that will have to be "terminated" because the machine malfunctioned, will not be realities within the very next years. Karyotyping is available now, as is genetic counseling, and questions such as "Shall everyone's genes be tested?" and "What are the tested people to be told?" and "How are they to be protected from abuses in the use of such findings?" are all very much in front of us, right now.

Also, it is possible to distinguish roughly between the more difficult, remote scientific developments and those we are about to achieve (e.g., we shall develop a computer that can play expert Chinese checkers long before we develop one that can master championship chess). Yet such progress is not completely predictable. A scientific breakthrough *could* suddenly force on us some of the remote issues. Thus, thinking ahead seems quite necessary.

Once a technique is known and widely used, it is difficult to remove. The time to review new technologies—if there is to be hope of curbing them or of channeling them toward some uses and away from others—is *before* they are introduced. Therefore, the time to decide if amniocentesis is to be used, not only for avoiding diseases but also for breeding, is *now*, before it is used for biological designing. Similarly, if *in vitro* experiments are to be limited to those leading to replanting the fertilized eggs in the *donor* mother, not in host mothers or artificial wombs, *now* is the time to formulate such a ban, not after babies are born to test tubes.

While I feel strongly that the best time to review a new technique is when it is first conceived—or at least before it is fully developed and widely used—I disagree with those who hold that once a new procedure or technology is available it will inevitably be applied and our opportunities for societal guidance will be lost. Social forces do play a role. There have been a score of techniques that have not been used, either because they threatened commercial interests or because a combination of fear and conscience held them in abeyance (e.g., nuclear weapons). Thus, in principle, *edit we can*.

Technologists like to push the argument to the opposite extreme and state that it is society that *determines* whether a new technique is used or ignored; therefore, their reasoning goes, they can develop new techniques to their hearts' content, without concern for the social consequences or moral issues.

The truth seems to me to lie somewhere between the notion of technological determinism and a forced march of history that no one can alter, and the voluntaristic notion that societies can direct such developments. If society could shape technology to its needs, we would have created a cancer cure a long time ago.

If technological inventions, once made feasible, would more or less force their way into society, we would have hybrids of man-apes, such as gorilla-man and chimpanzee-man, which, according to Dr.

Bruce Wallace, Professor of Genetics at Cornell University, Ithaca, New York, could be produced in the same way cattle are bred. [40] As I see it, while technology is neither wholly determined nor wholly determining, technologies do provide new temptations and new opportunities for use and abuse; and societies affect the rate and direction of technical developments and their applications. The two forces interact, so both must be taken into account, and both must discharge their responsibilities. Scientists and technologists cannot ignore the fact that they add to the burdens of a society already overwhelmed by the need to manage its fate and to sustain its moral fiber. Citizens and their representatives must invest more time and energy in assessing the new genetics.

It must be emphasized that, even if scientists do not act irresponsibly in promoting genetic engineering and hence may not need very much outside oversight, past experiences with food and drugs suggest that the business community badly requires careful regulation. The moment any of these new genetic procedures could be mass-produced and marketed, corporations would surely push them, as they do deodorants or laxatives. For instance, if sperm could be sold in capsules and used without a doctor's assistance, advertisements such as Order Your Next Blond from Us, and We'll Make Your Next Child 5 Inches Taller or Your Money Back, would certainly spring up in no time. And if a douche could insure that the next infant would be a boy or a girl, attempts to make people feel guilty if they do not have a "well-balanced family" would be on the air sooner than you can say "commercial." Therefore, a whole wing of the Federal Food and Drug Administration should be set up to scrutinize these new products before they are unleashed to commercial forces. While it is true that society's fashions affect the use of technologies, these fashions are not randomly or freely determined but, like the shape of our automobiles, gowns, or holidays, are carefully groomed by the business, advertising, and mass media forces. We should have a say *before* these forces promote the mass use of genetic healing or breeding, and concerns other than higher profit should decide whether these new techniques are going to be sold in the marketplace and passed through it into society.

Most genetic engineering techniques available as of now are expensive in financial, social, and psychic costs, and this fact is often used to argue that one need not fear that they will be used on a mass scale and thus have societal consequences, for better or worse. Indeed, so far most of the new procedures do require a visit to a physician's office (e.g., amniocentesis), and several entail abortion or artificial insemination, and therefore exact an economic and a psychic price. However, these matters are subject to change. Historically, most techniques tend to become less costly per unit and more socially acceptable in time. Sex choices could become as easy and as private as taking a douche or a pill. Some genetic tests are already inexpensive and quite widely used (e.g., PKU). Procedures once stigmatized are now quite acceptable—the use of contraceptive devices, for instance. At present even most Catholics plan their families. (By 1970 more than two-thirds of the female Catholic population in the United States practiced birth-control methods other than the Church-approved rhythm method.) [41] So one cannot rely on costs to keep the lid on the genetic Pandora's box, especially now that scientific developments are prying open its cover.

The fact is that there is quite a bit of genetic engineering in progress *now*, and more is being introduced each year. Our biological foundation is no longer subject simply to *natural* selection of the fittest or of any other kind. Other changes which we deliberately introduce have genetic consequences that must be taken into account. These are part of the genetic side effects of public policy that society and individuals have to consider. For instance, the level of radiation to which we are exposed as a result of X-rays, tests of nuclear weapons, certain pollutants, and nuclear reactors set up in population centers should be examined as to their effects on genetic mutations, most of which are not beneficial but harmful. Similarly, the enthusiasts of population control for affluent societies should take into account the effect on the genetic health of the next generation brought about by women having children later in life. Today the average age of our population is twenty-eight, but in a stabilized population, this figure might reach forty. And while at present, 30 percent of our population is under fifteen and those over sixty-five constitute only 10 percent, in a population of zero growth rate the proportion of people over sixty-five would be equal to that of those under fifteen.[42]

In addition, direct genetic intervention has rapidly increased. Genetic counseling units, which were first introduced in this country in the early 1940s, grew from thirteen such units in 1955 to almost one hundred eighty in 1969.[43] By 1971 the centers had increased in numbers to nearly two hundred.[44] While the earlier centers served generally self-referred clients and operated for the most part out of university zoology or genetics departments, most of the later clinics were more conveniently based in hospitals and medical centers, and served patients referred by medical professionals.[45]

Finally, new laws are being enacted; and new crash programs, whose aims are genetic intervention, are being funded by state governments. In 1971 the Massachusetts legislature passed a law requiring mass genetic screening: all school children must be tested for sickle-cell anemia. The program raised many questions concerning the adequacy of the test and the protection of the privacy of the subjects.[46] All deserve additional exploration, since other states are following suit, and mandatory testing in Massachusetts might well be best abolished.

When black leaders pointed out that "their" illness was being neglected,[47] Congress came out with a crash program to find a cure for sickle-cell anemia.[48] Expenditures were raised from one million to twenty-five million in 1973.[49] Soon Tay-Sachs, a Jewish disease, was made the focus of such a drive.[50] Then Cooley's anemia, or Cooley's disease, that afflicts people of Mediterranean descent.[51] For awhile it seemed there would be an "ethnic disease of the month" to receive crash treatment. Several scientists argued that this was not the way to attack genetic illnesses, and recommended that deeper analytic problems, shared by these and all other genetic illnesses, should be studied.[52] The advocates of the programs reported that "their" illness had specific problems which the general attack would not resolve, e.g., there is not yet a reliable test for identifying Cooley's disease. Someone qualified to do so must review these competitive claims; it cannot be done by hurried, untrained members of a Congressional committee.

Already more than fifteen hundred illnesses that are partially or wholly determined by genetic causes have been identified, and the numbers of the illnesses that can be treated or prevented increase rapidly. In short, there can no longer be any doubt that genetic developments have acquired a scope which requires our personal and public attention. It is foolish to use lack of demand, shortage of supply, or the disappearance of old-fashioned inhibitions as excuses for not making up our minds as to how far, how fast, and in what direction we wish to go in this field. And who—doctors, congressmen, scientists, or each individual—is to make those decisions?

As I was noting down my new questions, I soon realized that I was harking back to my old one: Who shall keep an eye on genetic researchers and practitioners? Make genetic policy? I had sorted out my feelings about the different purposes to which the new tools might be used; I satisfied myself that some could be utilized without necessarily using others; and I found that one must worry about these techniques, both because the scientists do have extra responsibilities to look after the consequences of their findings and because the new findings were actually being used. This left "only" one major question, that of authority and power. Who shall make the needed decisions? By now, the sun was rising over the rooftops of Paris. It was the first time since the night before I was to defend my Ph.D. thesis that I had worked into the morning. The question of who shall decide these matters of life and death, health and quality, would have to wait another day—more precisely, the day that was just dawning. I went to sleep for awhile.

Part III:
THE THIRD DAY

CHAPTER FIVE

The Right to Know, to Decide, to Consent, and to Donate

The Need for Guidance

By the time I made it to the UNESCO hall for the morning session of the third and final day of the meeting, the conference was deep into its first coffee break. A good part of what one picks up at an international meeting comes not from the formally presented, scientific papers but from the discussions which follow them—in the corridors, at dizzy cocktail parties, and, above all, during the coffee breaks. In many of these exchanges, I found new confirmation for my basic position that an effective assessment mechanism of the new genetic techniques was badly needed.

This morning over coffee, Hamburger, my favorite participant, mentioned that in some countries the death of a donor of organs to be used for transplants is certified only by the surgeons who are to use them and that blatant ethical violations have occurred. A doctor who joined us added that in the United States, pathologists often remove pituitary glands when doing autopsies and sell them *sub rosa*. A participant from Africa referred to a black market of organs in which body parts suitable for transplants were sold to the highest bidder.

If an international commission were operating, I suggested, it would recommend to the various governments that the sale of organs be forbidden. Such advice, however, requires additional deliberation, since, as someone interjected, organs are in very short supply, and paying for them is one way to get more. Maybe "flesh banks" are needed, but they should be run by respectable medical authorities, with two doctors other than the transplant specialist certifying the death of the donor. The commission might also decide that transplant organs should be made available to those who need them on the basis of criteria other than the recipient's ability to pay the organ bank. The general health of the afflicted person, the age, family situation, and other such criteria (the person's social value?) should be determined by some authoritative, representative body, not arbitrarily by a few persons, and surely not in the marketplace.

Another area in which the Paris conference served to reinforce my view that an ethics review board was badly needed concerned the experimentation with humans on which I heard so often from my fellow sociologist Prof. Bernard Barber. [1] Shortly before I had left for the conference, the newspapers had been full of outraged reports about experiments conducted on syphilitics in Tuskegee, Alabama. The experiment began in 1932 with about six hundred black men, mostly poor and uneducated, from an area which at the time had the highest syphilis rate in the nation. Though a cure later became available, it was denied to the subjects of this experiment in order to allow the study to complete its run. [2] In a 1958 study in Los Angeles, forty-five infants, children of poor, uneducated families, died when given chloramphenicol, administered to determine its effectiveness as an antibiotic. [3] This was done, although seven years earlier it had been established that the drug is highly toxic, because some researchers wanted to compare it to other drugs. [4] These researchers may well have felt, as Dr. James Wechsler of Columbia University suggested, that the usefulness of the drug, when used judiciously, probably outweighs its disadvantages. Who knows? Who rules?

In the field of neurophysiology, experiments involving the placement of electrodes into various parts of the brain have been made on subjects who have not granted informed consent—they are often being treated for psychosis and epilepsy—and who are not aware of the possibility of subsequent serious brain damage. [5] It was recently revealed that a University of Texas research team, in a 1956–57 experiment, deprived seventeen infants of nutriments recognized as being necessary to their development. [6] According to testimony given before a U.S. Senate subcommittee, a drug that has caused tumors in laboratory dogs has been injected in low-income women in Tennessee as a three-month contraceptive and in mental patients to regulate their menstruation. [7] At the conference, I learned that, despite these often reported abuses, only very few countries have statutes limiting experimentation on human subjects. This news hardly made me feel that an international forum promoting such limitations was superfluous.

Another urgent topic appropriate for such a commission was dealt with only briefly during the meeting; but it has often been debated in recent years: the definition of death and concomitant issues. What about the dilemmas created in the families of the patients kept alive by so-called "heroic efforts" once consciousness has been lost and cannot be restored? Still, in most countries, the required judgments are left almost completely to the discretion of the individual physician. The doctors don't have the benefit of any kind of guidelines emanating from systematic deliberation and based on public consent; therefore, the more progressive doctors lack the kind of support a respected commission could marshal for their new procedures, and public education on the matter is uncertain and diffuse. As a consequence, a new definition of death, legal as well as social, is evolving all too slowly and causing unnecessary tensions, grief, costs, and a deflection of resources from those who are curable [8]

Even less clear, and hence even more in need of clarification, are the matters which concern the definition of when life begins. The issue was raised during the discussion of Austin's work concerning lab-made fetuses and on other occasions when euthanasia of severely deformed infants was explored. Some doctors chose to define life as beginning with conception (a definition which would, of course, outlaw abortions even of mongoloid fetuses); others would wait until the examining pediatrician certified that the born child was alive (allowing them to declare a deformed infant "dead"); others preferred to recognize three, four and one-half, or six months of pregnancy as the time when life begins. The U.S. Supreme Court recently applied the term "viability" in the way some doctors do—that is, as the point at which the fetus can be maintained without a link to the mother's body. This court ruling comes close to a resolution of the kind a commission might have formed, except that since the court's deliberations were not backed up by a public debate, its resolution will be slow to gain widespread acceptance. The ruling also suffers because it was not preceded by a study by a group of experts, and it slipped up on a vital detail: Does "viability" include or exclude the use of machines? If machines are included, with the development of artificial wombs a fetus of any age will become viable once it is removed from the mother. This clearly is not an acceptable position. Thus the Supreme Court, expert in law but not in other fields, did not benefit from advice which would have come from a Health-Ethics Commission and which could well have acted here as a "friend of the court."

A commission could do more than define the point at which a fetus acquires the legal and moral status of a human being and at which its removal becomes murder rather than something akin to scraping off an unwanted wart. It could also help clarify the intricate matters involved in gaining "informed consent" for whatever the practitioner or researchers wish to do. Lejeune had raised this question with regard to medical experiments on fetuses on whose abortions the parents had already decided.* He was particularly enraged by the fact that some physicians perform experiments on these fetuses: "Recently, we had a case where some Russian astronauts were doomed to die in their capsule because their satellite was damaged beyond repair. [I recalled that the CIA was supposed to have monitored the tearful farewell Kosygin bade his doomed men.] Would you conduct experiments on them just because they were doomed?"

While at first I was shocked by the notion of experimenting on the doomed astronauts, I wondered if careful deliberations by an ethics board would have suggested otherwise. If such experiments might suggest ways to avoid future fatalities *and* if it had been established that the astronauts were fully informed about the experiments and volunteered to participate, would it be so wrong to proceed?

But what is the "informed consent" of an unborn child? It cannot volunteer or refuse to participate. Who, then, should decide? Its parents? It is far from evident that they are competent enough to make these kinds of decisions. British law, for example, refuses to leave this matter up to the parents and requires that the court approve any experiments on children, not just those performed on

---

*In April 1973 the National Institutes of Health, in effect, imposed a ban on such experiments in the United States.

fetuses. However, in many countries, children are not protected this way. Again, I did not believe that I or a group of people meeting for a weekend in Paris could or should set the guidelines for such complex matters. A thorough and public exploration was needed. Nothing illustrated this general point better than the next session of the conference, by far the most troubling one of all in three turbulent days.

Two Pediatricians Experiment on Babies

Doctors are usually guarded, especially in the company of non-doctors, when the dilemmas of their trade are discussed. Anyway, international meetings are not the most apt occasion for semiconfessionals. But the two young men who presented the last papers responded to some unknown chemistry of colleagueship and trust; or maybe, since they were away from their home bases, to the opportunity to let go; or maybe because they represent a new, more open, breed of doctors. They surely did open up.

The first to report was Dr. L. J. Dooren of the Department of Pediatrics, University Hospital, Leiden, Netherlands. You have to take in slowly what Dr. Dooren is saying, because otherwise his carefully neutral terms, reported here without the tremendous emotion he conveyed as he read them, may elude you. Dr. Dooren's specialty is pediatrics. He discussed the treatment of infants suffering from severe combined immunodeficiency by means of bone-marrow transplantation. These infants have inherited "a complete defect of humoral and cellular defense capacities," and hence are defenseless against attacks by bacteria or viruses. Until four years ago, almost all these infants died within the first year of their life "from severe bacterial, viral, fungal, and parasitic infections." Either both boys and girls born to parents with the defective gene die, each newborn having a chance of one in four; or only boys are affected, each newborn boy having a chance of one in two.

Dooren reports, without joy, that some of these infants can be now saved by transplanting bone-marrow cells from a donor. Why no joy? Because such a transplantation "will, in practically all cases, lead to lethal graft-versus-host disease unless the donor of the marrow is identical with the patient (as determined by leucocyte typing and mixed lymphocyte culture). If the infant has siblings, then each sibling has one chance in four of having such identity with the patient. Even in such combinations some risk for graft-versus-host disease is present." In short, while one out of four will live, three out of four will die, many from the very therapy administered.

One may say with a measure of aloofness, "Well, they are doomed to die anyway." But there is a difference between a death which cannot be averted and one which is medically induced, and between living out your first year and cutting even that time short. (How long the survivors survive is unclear at this stage because the experiments have not been going on for more than a few years.)

Moreover, the death caused by the reaction to the transplant is tormented, or as Dooren put it, "dehumanizing." When the procedure fails, the infants become violently sick, twisting first in mounting discomfort and nausea and then in excruciating pain as their incompetent bodies try desperately to shake off the alien intrusion. For most of them it is a foredoomed struggle; as their bodies gradually fail, death relieves them of the agonizing intervention which soured their blood.

If one judges the procedure as "tolerable" for siblings, when there is a 25 percent chance of success, what does one deem the intervention for those children who have no siblings, who are recipients of bone-marrow of nonsibling donors—children whose chances of survival are much lower? Dooren himself characterizes the outcome of the experiment for them as "extremely bad."

What does it mean, Dr. Dooren asked, for the parents to consent to such a transplant? Do they really understand what it means and can they anticipate the agony? Would they have consented to the transplant, Dooren wondered aloud, if they knew what would be likely to follow?

Aside from the problematic issue of the informed consent of the parents (and all of us) for such experiments on their children, questions are raised by the troubled consent of the *donors* of the bone marrow. These, mainly siblings of the afflicted infants, are quite often children themselves. And as Dooren points out, once the procedure is fully explained to a prospective donor, and even if he or she fears the general anesthesia, the fifty-odd punctures (to take the bone marrow), and the loss of blood entailed, the donor-child can nevertheless hardly refuse to participate. The child knows at this point that he or she is the only one who commands the ability to save the dying sibling's life. Therefore is the child, once approached, left with a real choice? Could a child ever overcome the ensuing guilt feelings if he or she refused? With the chances of success so low, should such a traumatic question ever be put before a young child?

But not everyone agrees that people, children included, should be spared difficult questions, or even necessary pain.

During the discussion, Moltmann, the theologian, pointed out that our hedonistic culture is intolerant of situations in which people make sacrifices. He wished we would be more accepting of grief, pain, guilt, and, above all, sacrifice. He observed: "My personal idea is that human life, personal life, is not epitomized by the homo-apotheticus who suffers nothing and is free from all pain; but by the homo-sympatheticus, the moral, the personally responsible life that can be affected by sorrow, by grief, and by pain. Our understanding of health derives from our understanding of what a human person is all about."

At another point, he added: "We followed too much the pursuit of happiness and endless life without pain at all. And so we took the dimension of sacrifice out of our life altogether. I do not think it is immoral to ask a member of the patient's family to sacrifice his kidney."

His point seemed to me well taken. Surely no one would wish unnecessarily to impose the burdens of such decisions on young children. But it does seem far-fetched to refrain from asking a child to make a life-saving donation that entails only minimal physical risk out of fear that if the child refuses, he or she will feel guilty. However others disagreed. Several psychiatrists considered it "criminal" to place the burden of such a decision on a child, believing that the resulting guilt might cripple a child for life.[9] I myself wondered if the effect of being considered as a donor would be so traumatic if the donor-child were psychically healthy to begin with, and couldn't a psychiatrist be brought in to help work out any fears attached to the donation itself or the decision not to donate?

Still, the questions stand. The issue cannot be settled by the conjecture of one psychiatrist or sociologist or theologian. What we need are some data on these questions. For example, how do donors react under different conditions? And we have to formulate collective guidelines, drawing on such data. Should a doctor put a child in the position of deciding whether or not to make sacrifices, even when it is unlikely that the recipient will benefit? And how old must a child be before he or she can be so approached and asked to make such a donation? (Dooren's youngest donor is all of seven years old.) And could, or should, the parents decide for young children? Anyone observing Dooren could not doubt that he wanted help, and wanted it badly, in sorting out these problems. We could hardly help him then and there, without data on the effects of donations on donors and without the necessary time, in a meeting in which the public was not involved. On the contrary, the chairman, examining the clock, thanked Dooren for his paper and moved us to the next one. Discussion was to follow the next report, which was about a related line of experiments.

The floor was taken by Dr. Theodore Fliedner, of the University of Ulm in West Germany, who used a rather different procedure to help children afflicted with basically the same illness. (There was a polite disagreement between Fliedner and Dooren about whether it was the same illness or whether what Fliedner was treating was a more moderate version in which some degree of immunological defenses are present.) Fliedner's methods avoid the transplant of bone marrow, with its potential severe effects on the recipients. Instead he circulates the infant's blood through a machine, where it is brought into contact with normal blood that contains some of the stem-cells, cells needed to build up immunological defenses. This, it is hypothesized, might stimulate the production of such cells within the afflicted child.

To guard against a reaction to the alien cells introduced through the machine, Fliedner finds it is best to isolate the infant and to "reduce or eliminate the bacterial flora of his intestinal tract and skin." Isolation is achieved by putting the babies into plastic containers for one year or more. This makes it less likely that they will be infected before their defenses develop (if defenses do develop) and also, it is hoped, makes them less resistant to the new, imported cells.

The results? "In our group, twin boys with a congenital immune-deficiency syndrome have been decontaminated from their microbial flora when they were seven weeks old and maintained in plastic isolation units for more than two point five years. [I could not help but repeat to myself: two and a half years.] During this time several attempts to reconstitute the immune apparatus by means of stem-cell transfusion (using the mother as a donor) and of thymus grafts have been made. No permanent take was observed. However, there was a gradual development of some immune competence that was encouraging enough to finally introduce a new microbial flora step by step. The two boys were able to be discharged from the isolators when they were about two point five years of age. They are now outside the isolator for one year. Although they show inter-mittent infection, they do quite well."

Questions rushed to mind: Was it sensible to keep a child from birth to age two and a half in a box—not to cure him, but only to reduce his infection rate? to eventually discharge him, but only to return him to the hospital frequently, to be kept alive for a number of years unknown? And as to the immediate purpose of the procedure which was "experimental"—that is, research—rather than a cure, did the parents truly understand what they were getting their children and themselves into?

Could a child develop normally in a plastic container? Fliedner reported that psychologists supervised the children to be sure that no ill effects of their prolonged confinement occurred and to work them out if they started to develop. How reliable were those psychological measures that showed "good general development of the children?" Other studies suggest that confining the movement of children is very debilitating. [10] To what extent could the special efforts Fliedner and his team obviously made turn two-and-a-half years of plastic swaddling into an experience sufficiently varied to avoid damaging a child for life? And was it wise that these psychologists were consultants to the team rather than part of an independent evaluative procedure?

As far as I was concerned—as part of the new, more regulated world we had to move into—whenever possible, the reviewers of a procedure should be kept as autonomous as possible, separate from those being reviewed, and such separation of powers should not be arranged on an ad-hoc basis, but should be part of what is ethically expected and required by the research institutions.

Fliedner himself did not disregard these issues, and spoke in favor of an autonomous review mechanism. He said many of these malignant diseases require an experimental therapy which is not yet a generally accepted procedure. Thus therapeutic approaches need to be reviewed continuously by appropriate clinical investigation committees, on the basis of accepted ethical rules as—for instance—those laid down in the declaration of the World Medical Association in 1964.

When Fliedner finished reading his paper, there was a long, heavy silence in the hall. Faces were drawn, as if all were lost in their thoughts of sorting out their feelings. When the discussion finally resumed, it moved slowly.

Hamburger raised a question of ethics. So far, of the 119 children treated by all those involved, at least 87 had died, and of the others, only in the case of 12 can one speak of successful treatment, and, possibly, the treatment method was advanced. He asked both experimenting doctors: "Do you propose to continue with these techniques or not? And if yes, why? since they have such terrible drawbacks?"

"Fliedner suggested that the donor problem in some marrow transplantation might be circumvented in the near future by taking stem-cells from the blood," Dooren answered. "That would be very helpful, except that as of now, it is not possible. So we have to do it the old way."

He then paused, as if in pain. Then, lowering his head like a football player about to stubbornly take on an opponent, he continued. "We see that one hundred nineteen patients have been transplanted and that, apart from some identical twin combinations, only twelve have a functioning graft with absence of primary disease; most died. You say: 'Well, that's all in the game; we have tried it, but they had a lethal disease [aplastic anemia, leukemia, etc.], and would have died anyway. So why not try?' But I don't think it is as simple as that."

No one could charge Dooren with being unaware of what he was doing or unconcerned by the ethical implications of his doings. "I do not know, of the eighty-one or more who died, how many of them have died from graft-versus-host disease. Let's say forty; let's say fifty—or thirty. But the lives of the twelve who lived have been paid for by the unbelievable and almost inhuman suffering of many, many patients."

In a low voice, he then asked all of us in the hall, all those within the reach of his question: "Has it been permissible? And can we go on or do we have to stop and wait for better methods?"

But there was no response. The question remained in Dooren's lap, and he reluctantly concluded: "For the moment my strategy is, if possible, not to treat a new patient before the clinical history of the last patient treated has been fully evaluated and discussed by a team of experts, including experts in the psychological field. I refuse to treat children so quickly one after another that this full evaluation will not be possible because of lack of time. In this way it has been possible to treat each succeeding patient a little bit better than the former one, and we have been able, somewhat, to alleviate several problems in each new patient. With each new patient we have to be very critical concerning the indication for transplantation, and very careful for the patient, the parents, the donor, and the members of the team."

Hamburger, with whom I felt more simpatico the more I heard him talk, again struck the right note: "Thank you, I think you gave the exact answer that we were all expecting to your question."

Hamburger's gesture was welcome. Dooren was in a hell of a spot; I would rather carry bricks or key-punch IBM cards than be in his shoes or in his lab. At the same time, I was not sure whether I should allow my compassion for Dooren as a person to cloud my judgment of his actions. Was he really coming up with a cure? Would it not be better for all concerned if a third, dispassionate party helped decide whether he should continue the "old" way or should wait for a while, at least until the merits of Fliedner's apparently more humane method were established?

Now Fliedner responded to Hamburger's earlier question: "I would like to tell you of an experience similar to Dr. Dooren's. The patient had a healthy brother who could, in principle, donate identical bone marrow. But we didn't do it because, clinically, within the isolator, the little one-year-old infant seemed to be healthy. Although the child was not actually healthy, a transplant of bone marrow in these circumstances would have meant that it was being done with the knowledge that I was giving to this seemingly healthy child cytotoxic therapy [which involves toxic effects on the cells] and bone-marrow grafts that would most likely result in a graft-versus-host disease. We wanted to see how the child managed, for we felt that a bone-marrow graft, if necessary, could be done later."

Instead of clarifying the issue, this further confused it. Was there a real chance of spontaneous generation of the immunological defenses, that is, without treatment for these poor kids? Or was

Fliedner, swayed by the delight of the parents playing with their "seemingly" healthy child, waiting to intervene only when the illness became visible?

Fraser put into words what I was thinking: "I'm sorry I can't come to any conclusion; I'm confused regarding a simple factual problem. It is axiomatic that in any medical treatment, the risk is weighed against the benefit, and after hearing the terrible catastrophies to which Dr. Dooren's treatment can lead, and after hearing on the other hand that Dr. Fliedner is more satisfied with the progress of the twin boys with the more conservative treatment, I wonder why Dr. Dooren didn't answer the question about the more conservative treatment. I would like to know his prognosis for these children if bone-marrow transplant has not been attempted. Is it a question of genetic heterogeneity, and are they discussing different diseases, or what is the situation?"

Dooren replied, "The patients discussed by Professor Fliedner and myself were suffering from severe combined immunodeficiency, that is, a total lack of humoral and cellular immune capacity. These infants with a total lack of immunological defense mechanisms generally die within their first year of life unless they can be immunologically reconstituted by bone-marrow transplantation. It seems, however, that there are also infants with a somewhat less severe form of combined immunodeficiency with some immunologic capacity left. I do not know if Professor Fliedner agrees with me, but I think that in his patients with combined immunodeficiency there was some immunological competence present, although it may have been extremely low. In such cases, one must perhaps choose to wait and hope for spontaneous improvement while the child is protected by reverse isolation, especially when no identical donor for transplantation is available."

Fraser pushed his point one more time. "Then you think that these are different diseases?"

Dooren, honest to the core, did not seek to evade the issue: "I do not know whether clinical forms of severe combined immunodeficiency that differ somewhat in the severity of the immunological defect can be considered different diseases."

It would be terrible enough to have a child born without immunological defenses. But if the doctors themselves cannot agree on these experimental treatments, how is a parent to choose between them? And do doctors tell their patients that there is a choice? Are Dooren's parents told about Fliedner's work? The issues raised by the kind of information provided to patients, the conditions under which they could provide "informed consent" to experimentalists and healers, pointed again to the need for a sage body, such as my pet Health-Ethics Commission, to lead public discussions, education, policy making, and debates on the new procedures that medical services will have to develop to handle these problems.

The Paris conference, of course, was not the first occasion on which questions have been raised about the right of patients to know what is being done to their bodies and to be consulted before action is taken. On the contrary, these questions have gained more and more attention in recent years, reflecting a generally increased awareness of institutional and professional authoritarianism, oppression of poor people, and consumer and civil-rights activism as well. The issues raised range from relatively slight ones—such as that of a patient frightened by the beep of a heart monitoring machine to which he was plugged, of a woman given medication by a doctor too rushed to explain its nature or expected side effects—to matters of great consequence.

One dramatic example of the general problem stuck in my mind. Only slightly less terrible than the question posed by the Dooren-Fliedner experiments ("My child will die before he reaches age one unless he is treated; but which of these new experimental treatments, if any, should he be submitted to?") is the question faced by many women when they are told that they need breast surgery when malignant lumps are found. Many surgeons choose what is called "radical mastectomy," removing not only the infected breast but, in order to reach cancer cells that may have spread, considerable surrounding tissue extending to the armpit. However, in recent years, evidence has mounted that in at least some categories of cancer, removal of the breast alone achieves approximately the same results. [11] Often patients are not told they have a choice. [12]

Dr. George Crile, Jr., of Cleveland Clinic, Ohio, has commented: "For too long surgeons have assumed the entire burden of deciding how patients with breast cancer should be treated. In the days when it was agreed that radical mastectomy was best, there was no alternative. Today there is no agreement, and therefore the surgeon is obligated to inform the patient of the facts. Only when the patient is allowed to participate in the decision can she accept an operation on her breast with what can be known ethically as informed consent." [13]

If women were told of the studies that show no difference in the chances for survival, it is unlikely that many would accept the deformity that goes with radical mastectomy. This extensive surgery, plus the radiation that usually follows, often leads to swelling of the arm, limitation of movement, and other undesirable "side effects."

It seemed to me quite appropriate that under these circumstances, when no clear medical preference is indicated, surgeons ought to let the *patient's* values form the decision. For example, a woman who builds her life around outdoor activities may treasure her capacity to move freely more than a woman who lives the life of a bookworm.

Also, such decisions become less abstract and, in a way, easier to make, when, in addition to data, we are provided with a slice of experience. Thus, when parents must decide if they are to abort a mongoloid fetus, instead of being obliged to act merely on the basis of statements made in their doctor's office, they should be given an opportunity to see mongoloid children in an institution, as well as to talk with some parents who keep their deformed children at home and love and cherish them. But such notions about a change in the proper conduct of physicians or, more generally, in the way people make decisions, are not implemented overnight following a flash of insight or good will. They require educational mechanisms (e.g., changes in what doctors are taught in medical schools), local review boards, and so on. And somebody has to concern himself with seeing to it that all this is brought about.

A question by Steinberg, addressed to Dooren, brought my attention back to the dialogue in the meeting hall. Steinberg had a knack for asking penetrating questions. "I may have misunderstood Professor Dooren—indeed I hope I did. But I think I heard him say that blood samples are taken from people, that the samples are typed for the HLA type [the many types of antigens whose development determines the success of tissue transplantation], and that this information is kept on file without these people knowing that they may eventually be asked to donate bone marrow or a kidney for a recipient. Did I hear correctly that this is done without the knowledge of the donor?"

"I'm answering on behalf of transplant organizations, although it should not be me, but people who are actually doing all this work, who should be answering," Dooren replied. "But I think I can say that most of these people are typed in the course of several programs and their types are kept in the computer. These people are not told where this may lead, nor that the computer may indicate them to be possible donors for bone-marrow transplantation. Is that an answer to your question?"

Steinberg quietly stated: "It is an answer to my question, and I regret that I did not misunderstand you. I think this is completely unethical."

These are strong words, rarely uttered in such meetings. All eyes were on Dooren, but he did not respond further. The discussion turned elsewhere, leaving me wondering: Should the patient's right to know and to consent be treated as an absolute one, even in cases like this?

It seems clearly desirable that when a person's blood is tested in a hospital or by a doctor or in the course of routine treatment, it should also be determined whether he or she could serve as a potential organ, blood, or bone-marrow donor. This information should then be kept in a computer, where it would be readily available. Often people die because appropriate donors cannot be found quickly enough. [14]

At the same time to inform everyone whose blood was typed that he or she might be called upon one day to donate this or that, would put on hundreds of thousands of people an extra psychological burden. Long before they are asked to donate anything, many people—most of whom will never be asked—may start worrying about how they might respond to such a request if it came from strangers or relatives, if it was for a vital organ during life or posthumously, and so forth. On the face of it, the answer seems obvious: Why not wait until an actual need for a decision to donate has arisen before informing them?

The assumption behind this suggestion—to "spare" the masses the burden of reflection, even of anxiety—is one that often underlies the position of those opposed to fully informing the patients. However, it is a highly paternalistic and patronizing view. People are seen as immature children to be protected by the doctors who know better and who will make the "tough" decisions for them, informing the laymen at the last possible moment. It seemed to me that people should be told and that the "psychological burden" means, in effect, that people will be given a longer time to think about the issues involved before they may have to make an actual decision. Why shouldn't people think—yes, even worry—about these matters? Isn't it at least as worthy a topic as any they would think about otherwise? (What will happen next to a character in a TV soap opera? Who will win the pennant race? What to cook for dinner? Should I change jobs?) Isn't our general tendency to deal with people as children one reason why they often act "immaturely"?

Moreover, if everybody is to be typed sooner or later, would not much of the onus of the decision be reduced and the burden shared by interpersonal and public discussion? And, could one possibly expect new moral standards that deal with these new issues to evolve without a great deal of dialogue, moral rethinking, and reformulation?

Take, for example, the great shortage of donors of organs.[15] Several doctors have suggested that people carry a card showing that they have dedicated their organs to the living, in the event they are involved in a fatal accident or die under conditions where their organs could still be used (e.g., die young enough, from natural causes). This suggestion, which runs counter to past traditions holding that the body be buried intact wherever possible, has not caught on. For one thing, people's natural anxiety will not be alleviated unless they can be assured that the death certificate will be signed by someone other than a transplant surgeon. As long as these matters are not openly discussed, or as long as there is no public debate and no mobilization of the moral forces of the community in support of such a new approach to organs, the old taboos will give way only very gradually.

For all these reasons, I feel that treating people as adults is the best way to help them be mature. I do realize that there are some arguments on the other side, at least for not telling all. When my father was dying from cancer, he didn't ask once—even after he came out from exploratory surgery—about what was wrong with him. Other people may show even more clearly that they do not wish to know what the problem is, and one should not force them to hear. But is it up to the doctor alone, or only in conjunction with the family or friends of the patient, to make the decision to tell or not to tell?

One cannot simply say that all doctors should, all the time, tell all the patients everything. There is a sizeable part of the population—some estimate it to be as high as 25 percent [16] —who are psychologically disoriented even before they are faced with crisis-decisions. Studies also suggest that when a major anxiety is added to an already critical mass, many "normal" people may suffer a breakdown. [17]

We need a transition period before the notion of "tell all" can be fully implemented. People will have to get used to their new responsibilities. Physicians will have to learn to share information and how best to share it. Clinics and hospitals will need more psychiatric nurses and social workers—or patients' group meetings—to help people to learn to cope with more demanding decisions.

There is no question in my mind about the basic direction in which we should move—we should bring people in on as many decisions as possible as soon as possible, as fully as possible. But this should not be done naively, with an untutored reformer's zeal, that will only lead to backlash and retreat. Opening up to the public, involving the citizen, should be combined with public education on the issues involved as well as institutional and professional reforms. These, in turn, require the leadership of those citizens who are already well-versed in these matters and who care about transforming one more area of authoritarianism and paternalism into one of authentic, well-informed participation.[18]

The meeting was breaking up. I had agreed to go out for dinner with Dr. John Case, but there was half an hour to kill. I recalled that one of the papers, handed out but not presented because the author did not appear, was by Dr. Henry Miller and dealt with related issues. I used the time to peruse it.

Spokesmen for the Doctor's Side

Dr. Miller tipped his hand early in his paper when he pointed out that ethical, social, and economic questions about the new forms of treatments "that we owe to the technological and scientific revolutions in medicine" and to new research procedures, were raised "especially by patients' organizations . . . sometimes supported by a small group of physicians who are unconcerned about and sometimes positively antagonistic to research." It seemed to me that while it was true that some of the questions emanated from these quarters, it was by no means true for all or even most of those who expressed concern. And in any event, whether the issues were raised by agitators or persons of good will, this did not establish the merit or fault of the points made. These must be judged on their own ground, rather than be challenged on the basis of the status of those who advanced them.

Miller's paper stated next: "The difficulties of the therapeutic pioneer can be exaggerated. The surgeon who tries out a new operative procedure does so because he hopes for a better therapeutic result, and while events may prove him to have been unduly optimistic, his good faith is rarely questioned."

Really? I wondered. There are several studies, at least in the United States, which reflect somewhat less kindly on the profession, pointing to overuse of all medical services, surgery included, whose motivation is, all too frequently, higher income for the doctors.[19]

Miller turned next to the use of new drugs. Here, he stated: "Often a doctor is quite likely to feel he should conduct a controlled therapeutic trial, in which the new treatment is measured against an established routine, or especially if there is no established routine treatment, against the effects of administering tablets of sugar or something similar. The ethics of the controlled trial have been the subject of some discussion. There are some distinguished British physicians who feel that while there is no need to warn the patient that he is being given a new drug in the ordinary course of clinical practice, it is dishonest to conduct a controlled trial without acquainting the patient with the fact that he is participating in it. The main reason for this objection is that it is regarded as introducing an element of deceit into the relationship between doctor and patient. I think most British physicians do, in fact, ask the patient to collaborate in therapeutic experiments of this kind, but personally I do not regard omission of such information as of any particular importance. Every course of treatment is an experimental trial, and other things being equal, a controlled trial is likely to be more carefully supervised by a more careful physician than the uncontrolled experiment of routine treatment."

I did not have to travel very far to find a doctor who doesn't see a great need to inform the subjects about what is to be done to them, and prefers instead to rely exclusively on those who conduct the experiments.

Miller went on: "There is to my mind only one real dilemma in this particular connection and it is exemplified by recent experience at my own hospital. A very careful five-year clinical trial, carried out concurrently in Newcastle and Edinburgh, has indicated that the administration of Clofibrate (Atromid S) seems to have a remarkable effect on the prognosis of angina. This effect does not seem to be related to blood cholesterol levels and the drug does not seem to improve the prospects of those who have had a cardiac infarction without angina. The results look convincing and suggest something like a fifty percent reduction in the expected number of further heart attacks during the period of observation. However, the history of the fiasco of routine anticoagulant treatment, where a tremendous amount of resources was wasted and a good deal of harm done by treatment which we now know to be virtually worthless, means that these observations must be carefully repeated elsewhere. The problem is quite simply how far these favorable results can be ignored by withholding Clofibrate from a control group of patients. Of course, if it were absolutely certain about the results of these two trials, the withholding of the drug would clearly be unethical. Some physicians already take this attitude at the present time and on the present evidence. However, even the best designed and executed experiments may yield fallacious results, and I consider it so important to get the answer right that further controlled trials are necessary to make absolutely certain. But here I am coming dangerously near another important ethical question—Is it fair even possibly to disadvantage any individual patient with a view to adding to knowledge which will help others?"

I felt several things at once. I continued to feel that one should never subject patients to an experiment without informing them that they are being experimented on and without securing their consent. I could see that the patients couldn't be told who got the medication and who the useless sugar pill; there would then be no valid study because the subjects' feelings could affect their responses. But they could all be told that they were subjects of a study and even that some would be given a pill and others a piece of sugar, and they could be told the reasons why the researchers couldn't reveal who got what.

Miller turned next from new treatment to basic research: "What about the critical area where investigations are undertaken not for the benefit of the patient himself, but to add to knowledge in the hope of helping future patients? A few doctors feel that this is never justified, but if it were abandoned, medical advance would virtually cease. I think two conditions should always be met in this connection. First, the procedure should be carefully explained to the patient and his permission obtained. Secondly, the doctor himself should be honestly convinced that the investigation is justified in that its probable contribution to medical knowledge outweighs the slight risk that is admittedly inseparable from any medical procedure."

Again I felt it was not enough to rely on the "honesty" of the research doctor. Some review by disinterested parties was called for, and beyond that, my feeling was that these needed some guidelines, carefully considered and not formulated solely by doctors. That abuses did occur, Miller fully noted: "One vexed question concerns the use of prison inmates or mental defectives for mass experimentation. So far as can be discovered, the several hundred mentally backward children at Willowbrook who have collaborated in valuable work on hepatitis which has led to advances that point the way to partial protection against this now dread disease, were virtually conscripted, though parental consent was obtained. Observations carried out in Britain on the effects of intrathecal tuberculin injection under somewhat similar circumstances some years ago raised a storm of protest, and the Willowbrook experiments have been criticized both in the United States and in Britain."

But Miller, again, found himself on the side of the physicians, observing: "We inoculate millions of children against poliomyelitis so that a small number of them do not develop the disease. Despite a recognized risk of serious complications, we vaccinate millions of people against smallpox, not so much for the benefit of each person vaccinated, but to protect society as a whole from the spread of the disease. When one considers the demands society makes from its members on very dubious moral grounds, especially in times of war, one sometimes feels that doctors are torturing themselves unnecessarily about such matters."

Ah, Dr. Miller, I said to myself, two wrongs do not make one right; if public authorities put dubious demands on citizens, this extends no license to the medical profession to add some of their own.

I must have been absorbed in my reading because I did not notice that Dr. Case, for whom I was waiting, had arrived. (Dr. Case is not his real name; the reasons for withholding it, while giving everybody else's real names in this book will soon become evident.) He was now standing next to me, saying, "Ready to go?"

We decided to look for an inexpensive restaurant that was crowded with Frenchmen rather than tourists. This way, we felt, we could not go wrong. Using these high-power sociological criteria, we ended up in Chez Maurice below Montmartre, a restaurant that qualified on both accounts. After the meal we walked through Place Pigalle and up the hill into Montmartre. It was far from an idle stroll.

"What are you doing in the States?" I asked.

"I head a genetic counseling clinic," he replied.

This was a chance not to be missed. Sociologists often study the process through which innovations make their way into existing practice. I asked him if he now routinely informed pregnant patients who were over forty about the availability of amniocentesis, the test to discover genetic defects in the fetus. (Over the last three days we had heard several reports by doctors from Western Europe and the United States suggesting that amniocentesis is highly reliable and entails only a low risk to the mother and, as far as is known, to the fetus.)[20]

To my surprise, Dr. Case asserted that he did not tell these patients about the availability of the test. Earlier in the conference, Lejeune, the French geneticist, had vehemently attacked the procedure on moral grounds, for if the test indicates that the fetus is defective, and the parents don't want it to be born, an abortion would be indicated. Lejeune, a devout Catholic, chose to define a fetus, even if only a few weeks old and still an unformed mass, as a live child. He called the procedure "murder" and suggested that science focus instead on developing procedures to cure those born deformed. All the doctors at the conference concurred that such cures are not available now and are, at best, a very long way off.

But Dr. Case was not a Catholic. Nor, I soon established, was he worried about the safety of the technique. (The test is known to have damaged at least one embryo, whose eye was penetrated by the needle,[21] and abortions in the fourth month of pregnancy, which the tests often indicate, are "late"; however, if both procedures are carried out by competent doctors, they involve only small risks.)

The reason this genetic expert (and others) chose not to inform his patients was, he said, to avoid generating "false worries." Scores of mothers with healthy fetuses would worry in vain until the test results were known, and some would continue to worry once the possibility of a deformed child had been raised, even after the test proved negative.

"Well," I wondered aloud, "I can see why you would not wish to alarm pregnant women unnecessarily, but in view of the grave consequences involved if the child is, say, mongoloid, isn't helping one parent to avoid a mongoloid child well worth the 'costs'?"

"Well," said the geneticist slowly, "I know of a patient advised to take the test who committed suicide. I cannot prove the test was the cause, but it might have been. I will recommend it only if asked."

Case's comment brought back to me the questions raised by doctors withholding information from patients or prospective parents. Should people be "spared," that is, protected, in this manner? Are doctors entitled to conclude that basic information about matters of such great consequence to the parents and the unborn child be withheld from them? Is there any evidence that most prospective parents cannot handle such information or that it causes anxiety more damaging than the agonies it might avoid? Can we leave it to each doctor to make these judgments alone, without any counsel?

The problems involved in the right to know and to decide what will be done (or not done) to our bodies have been raised systematically and effectively—although occasionally in a rather shrill manner—by Women's Liberation groups and publications.[22] They point to reports documenting many instances in which doctors have failed or refused to inform patients about what is done to them or what their condition is, of doctors who have disregarded their patients' preferences in matters which involve no true medical judgment, and who have refused to deal with the patients as persons.

Thus many women who express preference for natural childbirth report that most gynecologists will not agree to take them as patients or will pressure them to use drugs, although there is no *medical* indication that natural childbirth is less desirable; possibly it is just the opposite. While patients may face similar problems, doctors—mostly male—tend to view women as even less able to handle complex information and to think rationally; as a matter of fact, argue the feminists, physicians tend to see women as "hysterical" and inclined to "psychosomatic" illnesses.

I was wondering whether Case, like most doctors, would feel chiefly annoyed with the women's groups' protestations or would he recognize the legitimate issues they raised? But before I could articulate my question, we had reached our destination, the Métro stop, from which we were to be zoomed back to the conference hall. We could not talk much during the ride—although the Métro is not as noisy as the New York subway, the racket is loud enough. But on the way from our last stop to the hall, the conversation picked up. Case pointed out that the questions concerning the right to know also arise when artificial insemination is carried out. The mother is not told the identity of the sperm donor. But should she know something about his attributes? Many doctors, it is reported, use their own sperm or that of their students.[23] Should they first be required to undergo tests to determine whether they carry genetic illnesses in a recessive form? Should the prospective parents be told of the results? Should they also be told how often sperm from the same person is used in the same area so that they might take precautions to avoid the possibility that the various offspring of the sperm donor, biologically half-sisters and half-brothers, will intermarry?

A recent study suggests that the consequences of such unwitting incest might be quite considerable:

Of all the sexual taboos known to myth and anthropology, the prohibition against incest has the strongest clinical support, for the forthright reason that children born of such unions have long been known to have an unusually high rate of severe mental and physical defects. But until recently there were few if any scientifically controlled studies of the children of incest. Now, however, a Czechoslovakian researcher has completed such a study—and the results provide dramatic evidence that among the offspring of incestuous unions, the risk of abnormality is appalling.

Working through courts, hospitals, homes for unwed mothers, and orphanages, Dr. Eva Seemanova examined and kept records of 161 children born to women who had had sexual relations with their fathers, brothers, or sons. The same group of women also produced 95 children by men to whom they were not related; these half-brothers and half-sisters of the incestuous offspring formed Seemanova's control group.

The children of incestuous unions, she found, were often doomed from the start. Fifteen were stillborn or died within the first year of life; in the control group, only five children died during a comparable period. Among the children Seemanova examined at the Czechoslovakian Academy of Sciences, more than 40 percent of the incest group suffered from a variety of physical and mental defects, including severe mental retardation, dwarfism, heart and brain deformities, deaf-mutism, enlargement of the colon and urinary-tract abnormalities.

By contrast, Seemanova reports, none of the children born from nonincestuous unions showed any serious mental deficiencies, and only 4.5 percent had physical abnormalities. She thinks further studies of the children of incest are indicated but believes her data confirms the "unmistakable effect of inbreeding on infant mortality, congenital malformations, and intelligence level."[24]

Thus, persons living in a small town or city in which there is only one doctor who specializes in providing artificial insemination, and who decides to draw only on himself for sperm donations, could pay quite a considerable and rather wicked price.

All this is to illustrate—if it needs more illustration—that at this point most of these decisions, from the scope of surgery to be performed when breast cancer is found, to the safeguards undertaken in the donation of sperm, are now left almost exclusively to the discretion of the individual practitioner.[25] The best that patients can do is to consult more than one doctor in the hopes of finding one who is more willing than the others to share the information and the decision with them. But the reluctance of many doctors to do this is not accidental; hence most will be uncommunicative until a new consensus is formed in favor of more openness. Without it, those who would otherwise be inclined to open up, will fear censure from their colleagues as well as damage suits (as Case seemed to have feared).

As society needs both to protect the doctors from undue pressures so that they will feel free to take reasonable risks (which is essential, if new procedures are to be tried), and to protect patients from abuse, it might well make sense to bar malpractice suits, but to allow persons to file complaints against doctors before local health-ethics commissions. These would be more responsive to complaints because doctors would constitute only a minority position on them, and the commissions could, I suggest, bar a doctor from further practice. At the same time the commissions would eliminate greed as a motivation for malpractice. Moreover, insurance should be available to patients, rather than to doctors, to support them if they are seriously handicapped by faulty treatment.

Also, many doctors have an authoritarian proclivity and are reluctant to discuss matters in which their patients might want to have a say.[26] They prefer to pat the patient or parent on the shoulder, saying patronizingly, "Leave these matters to the good doctor," or even ". . . because I say so." Patients who can pay may buy a better treatment, but ward patients or lower-class ones will be completely at the mercy of usually white, male, middle- or upper-class physicians. If this is to be changed, counterforces will have to be generated to promote a new attitude. Most doctors will not change their bedside manners, at least not much, without rather considerable encouragement from their more progressive peers, the health authorities, ethical leaders, and, above all, the public. Nor do these matters simply concern the relationship of one doctor to one patient or parent. Nothing I learned in the conference better illustrates the larger forces at work than the discussion of the rise and the refusal to fall of the birth control pill and that of the question of how health priorities are set. That was next on the agenda.

CHAPTER SIX

The Birth Control Pill– Not for *My* Daughter

There was one more paper to go. This honor fell to Prof. Hilton A. Salhanick, who was pudgy, bespectacled, and vivacious, and whose titles, according to the program, were Hisaw Professor of Reproductive Physiology and Head of the Department of Population Sciences at Harvard University, Cambridge, Massachusetts. His was the only paper dealing with birth control; it started on a highly academic level and then quietly dropped two bombs.

Salhanick first reported to us, in the scholarly tradition, of all the difficulties he had had in defining his subject. How does one decide precisely when conception occurs? Is the birth control policy we discuss a matter of *family* planning or of *societal* planning? He continued in this same vein, enjoying the Socratic exercise of combining scholarly caution ("Words have many meanings; those must be cleared up before we can use them") with a mild sarcastic toughness ("The world is not what it seems to be; all is relative"). His tone of voice contrasted curiously with his intellectual mannerisms; he was reading his paper with gusto, as though he were serving up a large bowl of steaming spaghetti that he had just personally pulled out of a kettle of boiling water. He offered a taste of three criteria by which one could evaluate birth control:

"No currently known contraceptive method is ideal. The three primary considerations for a contraceptive are acceptability, effectiveness, and safety. Even the order of importance is flexible—it appears to me that the more enthusiastic the reviewer is about birth control in general, the more he emphasizes acceptability and effectiveness and the less concerned is he about safety. Conversely, those with antipathies towards birth control often exaggerate the dangers of a particular technique beyond probabilities."

Salhanick turned first to discussing how we can measure the acceptability of a contraceptive:

"First, the practice of contraception, quite apart from the method, may not be acceptable to some people. Thus, after five years of intensive publicity on family planning, over one-third of the married couples in the United States were *not using* contraception and the proportion *not using* contraception declined only one point two percent."

A slide was projected on a huge screen on the wall behind the speaker. It provided data in support of his statement.[1] Nonusers include some who wish to conceive, some who are sterile, and a third or less of the total nonusers who just take their chances.

Salhanick continued:

"There was a dramatic increase in the number of sterilizations, further emphasizing the lack of acceptability of available methods. Probably, the most critical demonstration of lack of acceptability is the high decrement rate, for all methods, about twenty-five percent per year, with somewhat higher rates in the first year and somewhat less after that."

"Acceptability" is usually viewed as the problem of the users: "Why don't they attend to their need for contraception?" or "Can't they accept some discipline?" or "If we could only convince the Catholic Church to drop its opposition to birth control. . . ." But it had already become quite clear to me that birth control techniques are as much at fault as the people, and tend to require a high degree of motivation and acceptance rather than accommodating the existing modes of behavior.

I had been made to realize this rather essential point when I attended a 1962 meeting at the headquarters of the Population Council in New York City. Here, as a member of a group of consultants, I was charged with examining the conditions under which people would be motivated to use contraceptives and to want smaller families. (The term used was introduced as "change in preferred family size.") We were then briefed about the various available contraceptive techniques and told not to worry about them but about motivation and values. However, during the discussion that followed, it became clear that the acceptability of a technique cannot be separated from its technical features. Pills have to be taken regularly; most people, especially in underdeveloped nations, do not respond well to such a regimen. Condoms require a high motivation and rely on the male, who is the less motivated of the pair. And so on.

It is also difficult to assume that technique and motivation can be separated, because acceptability (a motivational factor) and the two other factors, effectiveness and safety (technical factors), affect each other. That is, if a measure is not very reliable (a fair chance of getting pregnant still remains) and it isn't safe (e.g., it causes infection), it's no wonder the technique is not very popular.

A decade ago the intrauterine device was viewed by several leading experts as the ideal technique. Indeed, from a motivational and normative viewpoint, it surely qualifies. It can be inserted on one occasion, it requires almost no further attention; and it is always in place and ready for action. And people whose religion is violated can conveniently forget that they wear an IUD. But does it work? Far from perfectly. About 10 percent of the users of the IUD expel it during the first year after insertion, and many are unaware it is lost. It causes infection in 2 or 3 percent of the American population. Accidental perforation of the uterus also occurs, albeit rarely. [2]

Ten years later, in Paris, I found it startling that after all the investment, concern, and drive to curb population growth all over the world, technology still offers no simple, reliable device. It seems to me, from both a moral and practical viewpoint, that before one tries to change people's preferences, one should learn to service the values they already have. Actually, several studies suggest that if most unwanted pregnancies were avoided, most population problems would disappear. [3]

Unfortunately, as Salhanick's next sentences indicated, existing techniques are not fully acceptable, highly effective, or truly safe. Though Salhanick spoke in a diplomatic and academic tongue, his message is quite clear:

"Effectiveness has also been difficult to assess. *'Biological effectiveness'* in a small study group, carefully observed and continuously motivated, may be high; but in a large, random population, the same contraceptive technique might be found to have a low *'use effectiveness.'* In fact, in the absence of predictive behavior information, we are reduced to selecting contraceptives based upon environmental conditions and individual attitudes of the physician and patient rather than upon any form of scientific predetermination. . . ."

The fact that a technique must be very reliable to avoid unwanted pregnancies makes the job of finding effective procedures rather demanding. Thus one study of a group of couples estimated that, given a 99 percent effective technique—which is more protection than the best techniques provide—couples married for twenty years, who want three children, will have those within the first five years. In the next fifteen years 28 percent of the couples will get one "extra" child; 5 percent will be blessed with two, and a few will get twice the number they wished. [4]

Salhanick continued: "Now a few words about 'safety.' Unlike life and death, safety and morbidity are terms which have considerable flexibility. The use of the steroid agents is an important case in point. For the first time in the history of contraception, we include a morbidity and mortality rate in the evaluation of a contraceptive agent. We must ask, 'At what mortality rate should a contraceptive be declared unsafe?' Who should decide this? What allowance should be made for social benefit? With what condition should comparison be made—pregnancy? another contraceptive? sterilization?"

Salhanick was covering a lot of ground in a few lines, probably because he assumed that the people in the hall were already familiar with the data he mentioned in passing. But I was wishing that he would dwell on the figures somewhat longer, because much of the data are highly controversial.

The data Salhanick alluded to concerned birth control pills, which use steroid agents for contraceptive purposes. Years after the Pill was being used by millions of women, it was the subject of a flurry of critical reports suggesting that it caused a large variety of undesirable side effects, ranging from mild nausea to cancer and to blood clots, which, in turn, may cause fatal thrombosis. British studies have shown that the annual death rate due to lethal blood clots of Pill takers in the age group of twenty to thirty-four was 1 to 2 per 100,000; 3 to 4 per 100,000 in the thirty-five to forty-four age group. [5] The chances of being ill enough to require hospitalization because of Pill-caused blood clots are 1 in 2,000 for Pill-takers, ten times more than in non-Pill takers of comparable women... Other studies have made similar findings. [7]

One recent study, published in April 1973 in the *New England Journal of Medicine*, covered women between the ages of fifteen and forty-four in ninety-one different metropolitan hospitals. The study found that among users of the Pill, the risk of hemorrhagic stroke was twice as high than among the control groups of women who did not use it.

British studies show that when women on the Pill are compared with those who do not use it, the Pill takers are *nine times* more likely to be hospitalized for blood-clotting diseases and seven times more likely to die from such diseases (although the absolute rates for both groups of women are low). [8] Data on other side effects, from cancer to psychic irritation, are less well documented and more open to question than the data on blood clotting. Though there is evidence pointing to a correlation between taking the Pill and contracting some types of cancer, there are also some counterindications suggesting that the Pill users are less prone to develop cancer.

In turn, those data have been played down by some scientists. The development of the controversy illustrates the usefulness of hard data, rather than hypothetical models and arguments in debates on the merits of new techniques. The Pill controversy also points up the role that an authoritative "data court" could play in settling differences as to what is and what is not an effective and safe technique.

Since there was no such court to turn to, I relied on a book by two often-cited demographers, Leslie Aldrich Westoff and Charles F. Westoff. The conclusions in their *From Now to Zero* [9] are further supported by several articles I have read in *Science*. Data on these side effects of the Pill are reported to be worrisome enough for many doctors to take their patients off the Pill for various periods of time instead of keeping them on a continuous dosage. [10] Also, when the patient's family history includes one or more of a long list of illnesses, such as diabetes or kidney disease, most doctors recommend avoiding the Pill altogether. [11] But why is it prescribed at all? Why is it not taken off the market?

Two forces keep the Pill on the market and in millions of women's bloodstreams: companies that make a fortune from selling the Pill, and a theory that soothes the doctors. As to the market, the Food and Drug Administration was about to issue a stern warning—which was to detail all the dangers involved[12]—against the use of the Pill in March 1970. However, the drug manufacturers raised hell, [13] and the FDA modified the warning. Although an eight-hundred-word leaflet was made available to the doctors, the public was supplied with only a few lines of warning on each package of contraceptive pills. [14] The detailed warning was to have been distributed by the doctors to their patients. [15] However, no woman among my acquaintances has ever received such a pamphlet from her doctor, and Barbara Seaman, author of *Free and Female*, reports that the same holds true for the women she polled. She writes: "A year and a half later, I have failed to locate *one* pill user whose doctor has actually given her the pamphlet."[16] I have therefore provided, in Appendix 1 to this book, the original, undiluted statement the FDA had drafted.

The second reason no one is rushing to take the Pill off the market is that the data against it are not without some loopholes, loopholes wider even than those in the early reports about the dangers of cigarettes. Salhanick was explaining the difficulties involved in such studies:

"Unfortunately, in the case of the contraceptive steroids, we do not yet have a scientific answer to the inferences drawn from abnormal laboratory findings. Furthermore, it is disheartening to consider that we will probably never have an adequate answer unless a great number of effects should occur. I believe the opportunity to do a prospective study is long past. The lack of resources, the mobility of populations in countries which do have adequate financial resources, the difficulties in identifying causes of diseases, the variety and diversity of drugs currently in use as well as the low incidences of the diseases that are of concern, prohibit a definitive prospective study. Last, and probably of most importance, the ethical considerations of what is required—for example, continuous therapy of a large population over a long period of time by an impartial, dedicated team—are such that the proper experiments can probably never be done."

To put it briefly, Salhanick was saying that the Pill was "experimental" until a large number of women used it for a very long time. The dilemma is that it is hardly acceptable to prescribe the Pill for so many persons before it is safe; but you can never find out whether it *is* safe unless many take it—and even then, it is rather hard to tell. After all, we are dealing with only a few deaths per hundred thousand users, which does not sound like very much until your wife, lover, or daughter is the one who dies from it, or until you take into account that many millions are now swallowing it regularly and that the number of fatalities continues to rise. Is there no other way to get a better fix on how bad the Pill is?

At the very least, we should expect women to be better informed so they can decide if the Pill is worth the risk for them. Actually, the reports I read led me to conclude that women are not adequately informed. Aside from pressure from the manufacturers and "holes" in data, the problem exists because the data available against the Pill have been "discounted" with the help of a soothing theory. Many doctors believe in this theory and pass it on to worried women who ask them about the Pill. A review of this theorizing illustrates why it would be so useful to have matters like this examined systematically by a reviewing body instead of tolerating a condition in which policy makers, legislators, and physicians rely on such constructions.

When I was studying my notes after the Paris conference and decided to share them with the public through this book, I felt that many women—all those concerned with these matters—may wish to know more about the doubts raised in my mind when I hear the theoretical arguments as to why the Pill is the "best" technique available (a term used by the Westoffs and others).[17] I therefore provide here the main points made in favor of the Pill and the questions they raise.

Like many others, the Westoffs argue for the Pill as follows:

> Let us suppose that the eight to nine million women currently taking the pill discontinue. In an average year, after the situation stabilizes, what would be the change in mortality risks? Specifically, how would the reduction in risk of thromboembolism from the Pill be offset by the increase in mortality associated with the higher pregnancy rates resulting from the use of other less efficient methods?[18]

You might have noted a little assumption in the Westoffs's hypothetical situation—that women who will drop the Pill will get pregnant because, it is implied, they will turn to less reliable methods or use none at all. That assumption has yet to be explained. But let's hear the Westoffs out:

> Let us assume that all of the women in our illustration are trying to avoid pregnancy. First we need an estimate of which methods of contraception would be adopted in place of the Pill. Several studies have been done showing the methods couples used before or after the Pill and the methods they might choose if they were forced to abandon the Pill. On the basis of these different studies, a distribution was estimated which showed about 16 percent of the couples choosing each of the following methods: the IUD, diaphragm, condom, foam, and no method; 8 percent using the rhythm method; 3 percent withdrawal; and the remaining 9 percent using other methods.
>
> The next step was to calculate the number of pregnancies that would occur to these 8.5 million women using these methods compared to the number that would have occurred if they stayed on the Pill. The failure rates for each method is based on the 1965 National Fertility Study. The calculation revealed that an estimated 2.46 million pregnancies would occur to the 8.5 million women in the course of a year, while using other methods compared with the 340,000 that would occur while using the Pill. . . .
>
> The final step was to calculate the number of deaths that would occur as a result of thromboembolic disease incurred from the use of the Pill and the number of deaths that would occur from abortion and its aftermath. The results of such calculation indicate that 324 deaths would occur to the 8.5 million women on the Pill and that 1,179 deaths to the same number if they were using other methods of contraception. Thus, the risk of dying would seem to be three and a half times greater without the Pill.[19]

This theory—of course no one really knows what millions would do if the Pill were banished—reassures legislatures and public health officials and is told, in an abbreviated version, to many women who ask their doctors about press reports against the Pill. "It is safer than having a child," doctors often respond.

Whether or not you are soothed by this argument depends almost completely on whether or not you buy the assumptions. The strength of the Westoffs's theorizing is that it is not arbitrary but does draw on data, data on what some women did when they stopped using the Pill. It is assumed that if millions would stop using it, they would do what the small population that was studied did—that is, not learn a thing from their mistakes. Thus, just as 16 percent of the former users studied employed no other technique, so too would future millions, and so on.

If instead one assumes an active public educational campaign that directs former Pill users to the more reliable techniques—say, diaphragms—the figures *against* the Pill rapidly change. The same holds for serious efforts to teach women to use the non-Pill methods more reliably.

Also, the Westoff theory assumes that a full public recognition of the dangers of the Pill, up to its removal from the market entirely, would have no effect on the scope and force of the demand to develop a new, safer, and highly reliable preventive technique. That this is not the case can be seen from the fact that even now, when there is less concern with the problem than would exist if the Pill were to be banished, the quest for new devices is being intensified, although it surely could be further increased.

Let me briefly illustrate how readily one can come up with a set of rather different figures and conclusions. If we focus on women thirty-five or older, assume they would stop using the Pill and rely on local contraception, such as the condom or the diaphragm, and have no abortions, there will be 2.5 deaths for 100,000 women. If these women did back up local contraception with abortions, there would be only 0.4 deaths.[20] This compares with 3-4 fatalities, if they would rely on the Pill.[21]

The question of whether or not promotion of the Pill should be de-emphasized, banned, or continued is too important to be decided upon the spur of the moment or in a loose debate. Salhanick's paper again made me wish for an authoritative public body before which I could bring my arguments against the Pill and present my wish to promote greater reliance on diaphragms and greater investment in the development of new preventive techniques. The FDA, while tougher than the regulators of drugs in most other nations, is not strong enough, and worse, it is not a public authority. The FDA, making its decisions in bureaucratic secrecy, often timidly, and without open hearings or public debate, lacks public support for its decisions. But such support is needed if it is to prevail effectively against profit-seeking corporations.

Finally, it seemed to me, the right of citizens to know was again at stake. Whatever the government, scientists, or doctors might feel about the dangers of excessive population growth and the risks they might be willing to take to curb it, prospective parents should be given the facts, all the facts. The government may wish to influence their decisions by pointing out the undesirable consequences of large families for the nation and for those involved, but it shouldn't affect the prospective parents' judgments by withholding information about the dangers of the Pill or other contraceptive devices. Doctors should tell women the full truth. They can continue to tell women that the Pill is the most reliable technique, but they should also add that it is more likely to produce a fatal disease and bad side effects than several other contraceptive methods. Let those whose life and health are at stake decide what they prefer—a more reliable technique (the Pill) or a safer one (the diaphragm), and whether they prefer to achieve higher reliability (at the risk of dying from using the Pill), or if the diaphragm fails, to have a child they may not want (at the risk of death during delivery).

If the doctors feel that the rate of these dangers is low, they of course should say so. But all too often it is only on the doctor's authoritarian, paternalistic assumption that the decision is made for the patient, for, to reiterate, information on different illnesses and mortality rates apparently is not very often provided in doctor's offices, as my informal survey of friends and acquaintances quickly discovered.

Let us take a concrete example of what might be done:

Two doctors, authors of a widely circulated paperback full of self-help medical advice, say: "So much (maybe too much) has been said and written about the dangers of birth control pills. Let us try to offer some assurance. The dangers of the Pill are less than the dangers of pregnancy. . . ." [22]

What I feel they should say is this: If you would rather take a small risk to be absolutely sure you won't get pregnant, you may be wiser to use the Pill; if you don't want to get pregnant right now but don't mind too much if you do, you might use some other contraceptive means, e.g., the diaphragm. Aside from distinguishing between those who are and those who are not *very* anxious to avoid pregnancy, doctors should further distinguish between their young and middle-aged patients; they should not mechanically give the same advice to all patients. For older women the risks of the Pill are twice as high as for younger ones, and the danger of pregnancy is lower, if only because intercourse is less frequent. Also married, college-educated persons seem able to use diaphragms more reliably than unmarried or less educated persons. [23] Above all, the diaphragm does not interfere with the woman's hormone system, and if properly used, is quite reliable, although apparently less so than the Pill.

A doctor who favors the Pill read these pages before I sent them to the printer; he suggested that the dangers of blood clots could be reduced if women would have a check-up every six, or better, every three months, and be told to rush to the hospital if their legs swell, this being a common symptom of ensuing complications. This impractical and costly "solution" only reinforced my decision that I would not recommend the Pill to persons close to me. I therefore made a note to myself to see if, upon return to the United States, I could find a way to rekindle the discussions on these matters, which were in the public eye in 1970–71, but which burnt out before a conclusion was reached.

But to return to the conference: Salhanick discussed the complex ethical questions raised by the conflicting desires of families who may want children and a society that senses there are already too many. Then he returned to the Pill to illustrate the general issues. Without ever deviating from his almost jovial tone, he dropped another bomb:

"Our experience with the steroid contraceptives indicate that most of the serious side effects were unsuspected. Even dose-response data for contraception were not adequately evaluated, and to this day there is no known way to individualize dosages. Thus, assuming any reasonable sort of statistical distribution curve for persons on contraceptive steroids, almost certainly a large percentage must be ingesting more than is necessary for effective contraception."

Women who take pills with a high estrogen content are, it is reported, exposed to higher risk than those who take pills with a lower dosage. The FDA is encouraging the use of smaller dosages. [24] According to one study, when the amount of estrogen in the Pill used was reduced, the death rate of users was halved, from 3 to 1.5 per 100,000. [25] Another source indicates that high-estrogen pills (100 to 150 micrograms) are four times more illness-inducing than low estrogen ones (50 micrograms). [26] Britain is more effective than the United States in controlling commercial forces, and now allows only low-estrogen (below 50 micrograms of estrogen) pills to be marketed. [27]

Later I asked Salhanick why anyone would sell a Pill with a high dosage. He explained: "At first, the drug houses wanted to be certain of the antifertility effect, and they did not anticipate the thromboembolism and other potential hazardous effects. They were aware of the abnormal bleeding patterns with lower dosages. Consequently, they preferred to err on the side of antifertility activity. And they have slowly moved to lower dosages."[28]

My next question was a natural one: "Why slowly?"

"The reasons are complex and involve difficulties in evaluating preparations, getting approval by the FDA, changes in physician prescription habits, and not least of all, consumer resistance to change."

All this seemed quite true. But, surely, also involved here was the drive to make a profit, I felt. The Pill business is big business, and highly competitive. Given a choice between a few more blood clots (with higher dosages) and quite a few more pregnancies (with lower dosages) no firm, probably, would want to choose the latter and become known as the producer of unreliable pills. Indeed, a new "mini-Pill" is using no estrogen and less progestogen, about one-third or less that of the regular pill, but the risk of pregnancy is three percent as compared to less than one percent with the Pill.[29] And the FDA warns there is not enough data to show that the "mini-Pill" does not contribute to blood clotting. It seems a small percentage of the progestogen in the Pill is converted by the body into an estrogen, the hormone that can be illness-inducing.[30]

There was another difference Salhanick did not mention: the effects of overdosage were much less visible and traceable than those of a pregnancy. You *could* get blood clots some other way, but if you were pregnant, you knew where to place the blame.

The great surprise came during Salhanick's discussion of abortion:

"Abortion as a method of birth control presents unique ethical and emotional problems. I have tried to indicate that there are important differences between abortion for an individual with apparently insurmountable personal problems and abortion as national policy. The ethical debates, many of which hinge on the scientifically unanswerable question of when life starts, fall into three categories. First, there is the question of "destruction of life." The closer the abortion is to the time of conception, the less is the social ethical impact of the procedure. Alternatively, the closer to birth the termination of the pregnancy occurs, the more serious are the implications.

"Secondly, there is the important issue of a general disintegration of our society's respect for life. Obviously, those favoring abortion policies feel that the benefits of a liberal abortion policy for the living are more important than the interests of those to be aborted, while opponents cite life as an absolute value to be enhanced at any cost. Thirdly, and to my mind, most significantly, there is a modification of the lives of innocent persons. Women who undergo abortions have a higher incidence of premature children, and premature children are known to have a higher incidence of abnormalities as a consequence of the prematurity. Thus it is a matter of great ethical importance that we avoid bringing into the world humans with less than their full potential."

I felt that Salhanick had, in passing, dropped a bomb—much like a pedestrian casually tossing a kilo of nitro through the window of an armory. Most women do not, it seems, know that abortions have a deleterious effect on future pregnancies. On the contrary, most consider it an old wives' tale which has been disproven, at least in the case of abortions carried out by reputable doctors. I was so surprised by Salhanick's statement that, upon my return to New York, I checked with my research staff, all of whom are college-educated. None had ever heard about this anti-abortion finding. When I ran into one of the leading authorities on the subject, Dr. Christopher Tietze, at a West Side cocktail party, I asked him. It turned out that, if anything, Professor Salhanick had played down the findings in his carefully measured words. "A variety of British, Japanese, and Hungarian studies show that women who have had abortions are more likely to have premature babies," [31] Dr. Tietze said. "And premature babies are more illness-prone and death-prone than other babies."

The next day I got the British study (my command of Japanese and Hungarian leave much to be desired) and found in the June 10, 1972 issue of *Lancet*, an authoritative British medical journal, a report which plainly stated: "A tenfold increase in the incidence of second-trimester abortion has been demonstrated in pregnancies which followed vaginal termination of pregnancies." *Tenfold!* The study also referred to another report that supported this finding. [32]

Clearly, the public view of abortion underestimated the dangers involved.

I asked Dr. Tietze why the various anti-abortion groups have not publicized this finding.

"Probably because they are opposing abortion on moral rather than on technical grounds," he said.

This seemed to me a reasonable explanation, but also a sad one. The proponents of abortion tend to see themselves as humanists and libertarians, and to depict the opponents as religious fanatics, insensitive to the welfare of both individuals and society. However, this new evidence indicates that means of birth control other than abortion should be stressed. This is a case where the debate between the supporters and opponents of abortion has ignored a significant fact. Again I saw the need for a neutral, authoritative body to examine this new finding, and if it is further verified, to see to it that the public is appropriately educated concerning abortion, especially now that it is available on demand. It should be used to back up other birth control devices, not to replace them.

The discussion which followed Salhanick's paper seemed rather diffuse. One African representative asked about the ideal size of his country's population. Salhanick answered politely that he saw no way he could answer that. Someone suggested that if Salhanick was willing to make certain assumptions, he could provide an answer. I felt rather tired, and was wondering if I could sneak out and call home. I like to talk to my wife and sons at least once every few days when I am away. But just as I was trying to figure out what time it was in New York, I heard, dimly, a question which did intrigue me. Then Gellhorn, who was presiding, asked for my view, and I was involved again.

CHAPTER SEVEN

A Question of Priorities

Justice in Health Care

During most of the meeting, the discussion had been limited almost completely to the Round Table experts, who were in front of the hall. The ambassadors and representatives of international scientific organizations were seated in the back and rarely asked for the floor. But now an African representative, who sat too far back from me to allow me to read his name-tag, asked: "If I may return to the question which came to my mind earlier, may I ask how much it costs to try to save one infant in Professor Fliedner's bubble? How many African children could we save from simple, well-known illnesses, with these funds?"

The African's question was soon echoed by a white, leftist-leaning representative from an international organization who quoted the political educator Ivan Illich. Illich had written that "Latin American M.D.'s get training at the Hospital for Special Surgery in New York which they apply to only a few, while amoebic dysentery remains endemic in slums where 90 percent of the population live."[1] Illich added: "Every dollar spent in Latin America on doctors and hospitals costs a hundred lives. . . . Had each dollar been spent on providing safe drinking water, a hundred lives could have been saved."[2]

I had run into this viewpoint before. The American Friends Service Committee has noted in a report: "A single heart transplant costs $20,000 to $50,000. How many individuals could be rehabilitated with glasses, hearing aids, or dental care for the cost of one heart transplant?"[3] The same question was raised by Dr. René Dubos, a bacteriologist and scientific humanist at Rockefeller University, New York City, who asked: "Why become excited about a few hundred organ transplants when every day in New York, thirty thousand children are exposed to the possibility of permanent handicaps from lead poisoning and no one is doing anything about it?"[4]

Gellhorn, still presiding, commented: "I wonder why Professor Etzioni has not joined us yet on this issue."

I accepted the bait-invitation: "The major reason I am unusually reluctant to join the discussion is because this is a topic that deserves greater attention than we can devote to it today. I would like to briefly mention some of the complexities involved in this issue.

"Clearly there is a basis for concern with the societal dimensions of medical practice," I explained. "Together with my colleagues at the Center for Policy Research, I conducted a study of two hundred physicians in New York City. The study of cost-consciousness shows that most physicians are relatively unconcerned with the costs of treatment to the patient, less so when there is an insurance scheme, and least of all, when the costs can be charged to the government. Most doctors take the moral position that since they are entrusted with the health of their individual patients, no other considerations should enter. Even in areas where the public interest is directly that of health, such as reporting VD cases to authorities, many doctors do not go along. Three out of four gonorrhea cases treated by doctors are not reported, nor are all cases of syphilis—only ten percent are.[5] And when it comes to mere monies, the public perspective practically disappears.

"Now when we get into the question of allocation, here are some of the issues that crowd you," I continued. "We talk about the allocation of medical resources among different uses—for example, new research versus basic cures—but one standard answer is that if we are going to have one destroyer less, one less, then most research programs could be comfortably funded. So when we talk about allocations we not only have to ask about allocations *inside* medicine, but between medicine and other national services. Even if there is no change in defense spending, we spend eleven billion dollars a year on cigarettes. Surely one could argue that those monies could be spent more profitably.

"Beyond this, when the subject involves allocation of resources it quickly gets very nasty. It's getting nastier because the allocation systems of societies are not accidental; they are an integral part of the societal structure. Thus, in most societies, the more affluent get better medical services than the large middle groups and the working classes. This is not a slight oversight that can be readily corrected. The same is true for differences between the upper and middle classes. When people say 'Let's reallocate,' they are talking about very far-reaching social changes. Therefore even national health services, such as those in Israel, Britain, or Sweden, have not changed the fact that *money* buys the best service. And to bring about in the United States, for example, even the degree of egalitarianism that exists in the medical services of these countries, public perspectives, values, and power relations must change drastically. Nor will these be changed easily. Simple public medical insurance has been resisted in the United States for more than thirty years, long after it was available in many other countries.

"Now let's take just one more complexity, even more relevant than what was just discussed. The resource that is scarcest of all, and the one that cannot be extended by a stroke of a pen, is qualified medical manpower. The legislature of any country can move a billion dollars from any "priority" (say, defense) into another (say, health services) and the government can print another billion to buy a service.

"But what is very rare and what *cannot* be so increased is the kind of doctors we have sitting here—the talented, research-minded physicians who come in very small numbers. To begin with, it requires about fifteen years of training and enormous talent. Thus, when a doctor decides that he is going to work on illness X rather than on illness Y, it is not just the money being allocated that we have to consider, for the doctor is allocating himself—a much rarer resource.

"It may then be said, 'OK, he may work today on an esoteric topic, but tomorrow his findings may heal many persons with a large variety of diseases.' While there is an element of truth in this position, it is also true that when a doctor-researcher works on a rare illness rather than on a chemical element or on some other basic research topic, his findings are unlikely to apply readily to other illnesses. And we know, at least roughly, the incidence of various diseases. We have rankings of diseases in terms of fatalities they cause, days in hospitals, and days of other limitation of activities. In the United States, for instance, heart diseases rank first, and cancer, second, among the killing diseases, while gall bladder diseases rank eleven.[6] Hypertension kills thirteen thousand five hundred black Americans a year; sickle-cell anemia, three hundred forty.[7] The high-incidence illnesses may be much less interesting or much more difficult to solve; but if they affect millions, should they not be given priority? If you say, 'Sure, of course,' you must next ask: Can you tell the doctor what disease to study? Doctors usually choose 'their' illness because they have a research interest in it. How, if at all, can one administer their intellectual curiosity?"

I concluded by noting that this was not the time to answer all these questions. "All I wish to suggest is that the issue of allocation is indeed an important and very complex one."

During the break which followed, Dr. Alexander Bearn, Chairman of the Department of Medicine, Cornell University Medical College, got rather angry with me when I inquired if, since many parents who have mongoloid children simply dump them on the public, parents should be *asked*, when a fetus is determined to be deformed, whether they plan to take care of it themselves.

"You are really getting deeper and deeper under their skins with your questions," Madam Herzog said to me in the corridor. I was uncertain whether she got a charge out of my persistent queries, whether she was mildly aghast, or completely ambivalent. I thought I'd been rather gentle; after all, I could have said many other things. I could have said, for instance, that these doctors were wasting their talent and our monies. But it soon became evident that when I asked if doctors should be guided by anything other than their own light, many felt I stepped on more toes than I had realized were around. The UNESCO official, who two days earlier had enthusiastically asked for my support, passed me by with a formal nod. My two lunch mates of the day before grew distant. Lamy seemed rather preoccupied when I ran into him on the way out of the hall, and Klein capped it all by saying, "You're something else!" I avoided more "compliments" by returning to my seat before the meeting reconvened. I used the time to finish reading a paper by Dr. Henry Miller, which concluded with a comment relevant to today's proceedings, since it focused on the problem of limited resources:

A serious problem arises where resources are limited, but this is a situation that doctors have always had to live with. And while we all worry about our inability to furnish dialysis and transplantation to all young sufferers from chronic kidney diseases, and while the doctor must do his best to persuade society to meet his patients' needs, in the last resort he must do the best he can with whatever resources he has. The same applies with regard to the selection of cases for expensive or elaborate treatment. The best person to make this decision is the physician in charge of the case. He will usually be guided chiefly by the clinical prospects of benefit, but it would be idle to pretend that, whether implicitly or explicitly, questions of social usefulness and age do not play a part in his decision. To suggest that they should not do so is to deny common sense.

These few lines contain a progressive statement concerning the *political* responsibility of the doctor who ought to mobilize the society for more support rather than accept the status quo as satisfactory. But Miller's conclusion contains also a reaffirmation of the traditional view that the individual practitioner knows best and hence he should render the ultimate decisions, and an admission that, in making up his mind, matters other than the welfare of the individual patient are being considered.

As a matter of fact, though, most doctors do not recognize, let alone live up to, their political responsibility. In particular, they very rarely speak up or act on behalf of environmental causes (e.g., against pollution), consumerism (e.g., to keep out of food those additives suspected of producing cancer), the advancement of occupational safety (e.g., against the use of asbestos in construction), or other public matters, especially the one that should be very important to them—preventive health care. Yet it is precisely here where large gains could be made—before anyone got sick.

Most doctors tend to make, on their own, decisions on matters in which their patients or their family should be involved (e.g., weighing how much risk a patient runs of spending his or her last days as a vegetable because of excessive surgery, or which radioactive or drug treatments a patient should be exposed to in order to reduce the probability of cancer reasserting itself). And there is no reason why the individual doctor's notion of societal utility should overrule that of the community itself. I rather suspect that for quite a few M.D.s a movie star is more valuable than a professor, a "breadwinner" more valuable than a childless housewife, and a white middle-class person like themselves more valuable than most others. Hence there is a real need for the community to formulate its preferences, as is done on committees that decide who will get kidney dialysis. Unfortunately, this is done in practically no other area of allocative health-service decisions.

Moreover, Miller himself depicts what is really happening—the individual practitioner's views prevail on medical *as well as* on other aspects of the decision. He captures well the ideology of the medical profession in this matter:

> One hopes that the attitude of society to medicine will ensure increasing resources to extend the benefits of modern treatments to as many as possible of those who need them, and one also hopes that an improved climate of opinion, both public and professional, will sharpen discrimination and lead to heroic treatments being less often applied in hopeless or inappropriate circumstances. Gradually increasing experience of the implications of high technology in therapeutics is bound to influence practice. However, it is questionable whether physicians and surgeons who are actually looking after gravely ill patients pay much attention at that moment to the long-term effect of such activities, multiplied many times over, on society and the economy. Fortunately for the patient—and this applies even if he is a paralyzed infant or a demented old lady—the physician's professional instincts allow only one course open to him, which is to do the best he possibly can for the patient before him with the resources at his disposal. As potential patients I am sure we should not discourage the maintenance of this traditional attitude, and while we must discuss and consider the many problems raised by the advance of scientific technology in medical treatment, we should hope that the final decision about the individual patient will continue to be taken on clinical rather than on social or epidemiological grounds.

On reading this one cannot but think, at first blush, "Yes, indeed, if I were a demented old lady, I would still like to be given first priority." But on second thought, one says, "If only one heart were available for transplant, should it be given to me or to a young person? Could I still demand or expect top priority?" Even if you say, "Well, don't expect me to be noble when it comes to questions of life and death; I see myself (or my child) as coming first," the answer must be: "This is precisely the reason why allocative decisions are best made, not from the viewpoint of the concerned individuals *or* their advocates *or* the medical practitioner, but from a broader perspective." Social considerations should enter via such mechanisms as legislation (which would determine whether more funds should go toward the treatment of children or of the aged, to research on cancer or sex-change surgery, etc., etc.), and committees composed of doctors, theologians, and elected citizens. Such committees should decide who makes the decision as to who gets access to scarce resources, and they should draw on community values and public discussion rather than on individual preferences.

Thus the issues raised by the questions from the floor about health priorities and by Miller's lines have far-reaching implications. In effect, they ask about the totality of the societal composition, the place of the health system within it, and how both society and its health practices could be changed to bring them closer to our values.

Once more the conference opened a topic which required much thought, study, and action. I should leave it for now, yet I cannot just let it ride. At least the direction in which I feel we ought to proceed must be indicated.

While I am quite aware that the American health distributive system and priorities reflect the particular social, economic, and political structure of contemporary United States (as do those of other countries) rather than a highly rational, just, and systematic health policy, I still do not see it the way it is depicted in a recent spate of radical books. These attack the American health "empire" as one more example of an imperialistic, corporate, bureaucratic regime, run by a well-entrenched elite that systematically exploits the masses and services the few. I find these books difficult to read, because they constitute a crazy mix of well-taken criticism and irresponsible overgeneralizations; they are quick to attribute motives and slow to recognize the constraints of reality. Therefore one must read with special interest Edward Kennedy's book on the subject.[8]   After all, he is a U.S. Senator and a serious candidate for the presidency, and his book is backed up by considerable research work as well as evidence gathered in long Senate hearings.

Kennedy cites a well-known study by Dr. Ray E. Trussell, which found that of the cases studied, one out of five hospital admissions was unnecessary; twenty out of sixty hysterectomies did not have to be performed, whereas six others were highly questionable; one out of five hospitalized patients received poor care, and another one out of five, only fair care.[9] Dr. Lowell E. Bellin, who studied dental care, reported that 9 percent of the care was very bad; 9 percent, outright fraud; 25 percent, clear "overutilization" (that is, unnecessary service).[10]

And it is all big business. Kennedy points out:

> America's $17 billion-a-year health-insurance industry takes enormous salaries, commissions, and profits out of the premiums you pay, and does little or nothing to control physicians' and hospitals' charges or stimulate them to deliver better health care to Americans.

Kennedy goes on to depict a system of dealers rather than healers, concluding:

> I am shocked to find that we in America have created a health care system that can be so callous to human suffering, so intent on high salaries and profits, and so unconcerned with the needs of our people. American families, regardless of income, are offered health care of uncertain quality, at inflated prices, and at a time and in a manner and a place more suited to the convenience and profit of the doctor and the hospital than the needs of the patient. Our system especially victimizes Americans whose age, health, or low income leaves them less able to fight their way into the health system.[11]

One cannot but conclude that our health system must be rather seriously distorted if it leads a person as liberal as Kennedy to take so radical a stance. Also, though one may not be very outraged by commercialization of highways and holidays, the fact remains that when it comes to our very existence, we expect injustice to retreat. When our basic rights to life and health are obviously neglected, only the insensitive can remain indifferent and moderate.

Obviously my favorite commission-to-be—even backed by a wide range of supporters, and seeking basic health reforms and changes in societal priorities—could not eliminate many of these problems. It could, though, act as an agency to educate and mobilize public opinion around some of these matters. It could look into alternative arrangements and point to medical systems which are more humane, just, and responsive. Above all, it could serve as a symbol of what we need: *greater emphasis on systematic, publicized overviews of our health system and the mobilization of necessary social forces to enforce the superiority of human needs over the interest of service providers—be they industries, practitioners, or scientists—and, finally, the advancement of a decent, egalitarian, humane system.*

In contemporary society there are often so many issues in so many areas (genetics is just one; and health priorities, another) that we find ourselves paralyzed. It took a generation to overcome obsolescent laws about abortion. We still have not digested the data on the nonaddictive and probably harmless nature of marijuana. Even the enactment of a constitutional amendment securing equal rights for women is dragging on. As our society faces ever more challenges —from technology and the deteriorating environment to newly awakening social groups—we must learn to make more decisions more expeditiously, more wisely, and more fairly. Otherwise we will be hopelessly stymied.

The commission would be a good beginning, a legitimate point around which we could mobilize greater societal responsibility and responsiveness (hardly a cure-all, I know, but we must start some place). The concern with new genetics, the decisions about the engineering of life, of death, and of breeding seem good rallying points.

A national or international commission needs to be accompanied, preceded or followed by thousands of local ones. All communities should have review committees to help professionals—and to supervise them—in matters ranging from experimenting with human subjects, to deciding who will get what organs; from health priorities, to freedom to die with dignity. These are needed not only because the national and international guidelines must be implemented by thousands of individualized local decisions, but also because such local committees are the best way to involve large numbers of persons throughout the land in discussing—and thus becoming educated and active—in these matters. They can no longer be dealt with in an exclusive, monopolized, closed-door way.

The Commission Is Conceived

During the closing session, Hamburger leaned over a row of chairs to ask me if I would formulate a resolution, for the Round

Table to endorse, calling for the establishment of an international commission to explore further the ethical issues we had touched upon during the meeting. He added that he would translate it into French and co-endorse it.

I noted on a pad:

The CIOMS Round Table Conference, which met in Paris at UNESCO House on September 4 to 6, 1972, discussed the social and ethical implications of progress in some areas of biology and medicine. The conference recommends that CIOMS and WHO should explore the possibility of establishing an international body to explore and study the moral and social issues raised by new and forthcoming developments in biology and medicine.

Such a body would include biological, medical, and social scientists; humanists; religious leaders; and science policy makers; and will be backed up by a research staff.

I knew the conference had no authority to "resolve" and it would be futile—indeed, probably counterproductive—to try. So I'd written "the conference recommends." I passed the note to Hamburger, and he nodded his approval.

During the next break, I asked Gellhorn whether the resolution could be considered. He examined it, said he liked it, and suggested that I introduce it. I said I would consider doing so, but that I really felt the resolution would fare better if someone else were to introduce it. While I would have loved to have had the resolution named after me, I felt the issues it involved were so important that I should not risk jeopardizing them by introducing the resolution myself. I was one of the few non-M.D.'s in the group, and my prodding of M.D. sensibilities had not exactly made me the center of popularity at the meetings. Hamburger was similarly outspoken, and seemed quite willing to co-sponsor the resolution, but was he the best person to introduce it? So, as the conference was settling down to renew its work, I asked Gellhorn if he would introduce the resolution "for all of us." He wisely allowed a staff member to edit it first and asked the head of the UNESCO Science Policy Division to review it before he presented it to the assembly.

When the revised resolution came back, the phrase "nongovernmental" was inserted before the reference to an international commission, apparently to allay any fears that doctors may have of possible legislative attempts to curb their rights, as well as to reassure the governments represented at the conference that we weren't setting up a piece of a world government.

A paragraph was also added to placate the Austins, those proponents of the science-is-not-to-be-curbed viewpoint: "The problem facing the participants revolved around the possible misapplication of the results of some types of biological research and the responsibility of the researcher to society."

Application, it was implied, is where the problem lies.

Copies of the resolution were put before all participants. Now a collective editing of the text began. Numerous alterations in wording were made. Most of the changes suggested from the floor were rather minor. For instance, it was agreed that the words "as a minimum" would be added to the list of suggested participants on the commission so that no one would feel excluded; otherwise, it was felt, a full enumeration of potential participants would have to be made at this stage.

I decided not to oppose any of the changes. While I felt that the paragraph added for the Austins did not quite express the dangers posed by some of the new basic research, the wording was ambiguous enough for me to live with. At least it did not explicitly rule out basic research as a source of trouble. Moreover, I realized that at this stage the specific wording did not matter; the main point was to get the resolution passed. It had a long way to travel before the various international and national authorities accepted it, provided the funds for it, and put it into operation. If its trip was to be completed in the process, the format would have to change several times over. Therefore it was premature to argue about phraseology. Soon Gellhorn was stressing that the resolution was merely a recommendation; then still sounding much in favor of the resolution, he asked for the vote. In the final tally there were no Nays and only one abstention. The text of the resolution, as it was actually endorsed, is as follows:

COUNCIL FOR INTERNATIONAL ORGANIZATIONS
OF MEDICAL SCIENCES
CONSEIL DES ORGANISATIONS INTERNATIONALES
DES SCIENCES MEDICALES

ROUND TABLE CONFERENCE: "*RECENT PROGRESS IN*

BIOLOGY AND MEDICINE—ITS SOCIAL AND ETHICAL IMPLICATIONS"

Paris, September 4-6, 1972

RESOLUTION

The CIOMS Round Table Conference, which met in Paris at UNESCO House on September 4 to 6, 1972, discussed the social and ethical implications of progress in some areas of biology and medicine.

The participants of the conference, representing the biological and medical disciplines as well as the social, philosophical, and theological disciplines, examined the progress made in biology, its applications to medicine, and the development of modern techniques resulting from the said progress.

The problem facing the participants revolved around the possible misapplications of the results of some types of basic biological research, and the responsibility of the researcher to society.

It was felt that at present no mechanism exists which would enable individual practitioners, governments, and other policy-making bodies to arrive at decisions based on full knowledge of the facts and the moral, social, and ethical implications of their decisions, nor were there satisfactory criteria in view of which such decisions could be reached.

In view of the above, the conference recommends that:

1. CIOMS and its parent organizations, UNESCO and WHO, in conjunction with other national and international bodies concerned about the subject, should explore the possibilities of establishing an international nongovernmental body to explore and study the moral and social issues raised by new and forthcoming developments in biology and medicine.

2. Such an organ would include, as a minimum, biological, medical, and social scientists; humanists; religious leaders; science policy makers.

3. This body should be backed by the possibility of initiating and promoting research in the applications of biological and medical discoveries and their impact on society.

This, of course, is only a first step. It will take time, effort, and monies to turn the resolution into an alive and kicking commission. In turn, the commission will have a devil of a time making up its mind, as there could hardly be more complex issues than the ones with which it will have to deal. Also, as the commission will only advocate rather than legislate whatever guidelines it comes up with, the impact on national policies and individual practitioners will inevitably be gradual.

Nevertheless I was, and still am, quite sure that there is no other way. We are taking nature into our hands. We are turning matters once decided for us by forces we neither understood nor controlled into matters of human choice. The advances in genetics, both at hand and forthcoming, leave no alternative. We must now make decisions. Thus, since the advent of amniocentesis, it is no longer a verdict of nature that a given mother will have a mongoloid child. If one is born now, it is the result of a decision made by a doctor (Was the test made? Was the mother told? Was an abortion available?) and, one hopes, the prospective parents. Similar decisions have been made possible by the advent of other new techniques, (from *in vitro* procedures to karyotyping) and the new application of older ones (especially of artificial insemination).

And, if decide we must, the next question is what the relative power of the various parties to the decision will be. Who will make the ultimate decisions, and who will advise, consent, or constrain? The parents? the doctor? the law? Should there be an FDA for medical procedures—that is, to review doctors' new tools and inter-ventions—as there is now one that checks the drug and food manufacturers? Or can one rely on individual doctors and peer reviews to systematically follow up thousands of cases for years—as so often is necessary to fully assess a new medical procedure? However one answers these questions, the new decisions we must make are both so intricate and so far-reaching in their consequences that all parties will benefit from a systematic review conducted by one or more commissions, of the data relevant to the expected consequences, as well as from an analysis of the ethical and social and personal alternatives. Decisions reached through such a process *have* to be better than those made under the pressure of a frightened parent, the busy schedule of a physician, or the ad-hoc ways of politically sensitive lawmakers. Admittedly, an ethics board will not resolve once and for all many of the questions I've tried to outline, but it may well make dealing with them somewhat less difficult.

On the plane going home, I wondered whether I would dare have another child after all I'd heard. How could anyone? I thought how fortunate I was to be a Ph.D., not an M.D., and certainly not a genetic counselor, juggling such fateful decisions. More than ever before I was convinced that these questions were too vital to be left only to the discretion of doctors and their underinformed or ill-informed patients. If war is too important to be left to the professionals, [12] surely we cannot delegate decisions concerning our very lives to others who often do not have even our informed consent, and who at best are mindful only of our individual well-being but not of our collective life, which in turn forms the world in which we all have to live or die.

Postscript: The Long Road to Social Change

Beyond a Resolution

Those concerned with social reforms and change must be careful not to confuse the formulation of an appealing idea with the initiation of a real change. Ideas are common; effective reforms are rare and require long, sustained labor. A resolution passed at a convention may begin a process of social change, but only if the advocates of ideas and resolutions stay with them, marshal support for them, and face the opposition—which almost inevitably arises—will the ideas and resolutions have a lasting impact.

Before the Paris resolution could be turned into a worldwide and potent commission that would crown a hierarchy of national and local ones, the World Health Organization and UNESCO were required to act. Unfortunately neither body is spirited, innovative, or expeditious. For an American commission to come into being, the House of Representatives still had to add its OK to that of the Senate, which at that time—the end of the summer of 1972—was simply not in sight. The House, it was said, was preoccupied with other matters. Nor could the effort end even if the act were approved. Years of work would be needed in order to recruit the staff of the commission, to implement its programs, to set up local boards, and to develop their review systems. Creation of a more effective and more responsive societal guidance mechanism, of which the health-ethic review system would be but a forerunner, requires a continuous effort, and it is a task which has no foreseeable conclusion. Setting up the commission, though, seemed a good place to start; so upon my return from Paris, I tried to see if my colleagues and I could get things moving.

One obvious way to get action is to arouse public interest in, and demand for, the needed changes. But, as anyone who has tried to mobilize the public knows, it is not easy to rise above the general noise level and be heard. The press, radio, and television showed great interest in the breakthroughs in the fields of genetics and medicine—for instance, in test-tube babies—but they indicated almost no interest in the need for overseeing mechanisms. The Paris resolution was reported only in *Science*, and the death, shortly after, of the Mondale bill, aimed at setting up an American commission, was not even noted by the *New York Times*.

Next, and more discouraging, the Paris resolution was put aside by those who had to enact it. Dr. Gellhorn, back at his duties as the Dean of the University of Pennsylvania Medical School wrote in a letter:

> At the Executive Committee meeting of CIOMS, after the Round Table, the resolution to create a combined sciences deliberative body was considered. Both WHO and UNESCO shied away from direct monetary support of such a venture because of the internationally sensitive issues which such a group might consider. The United Nations agencies would be delighted to have the results of discussions by such a "commission," but could not be identified with its creation or direct support.

This outcome was hardly surprising, but it was disappointing nevertheless. Gellhorn went on to point out that, rather than a permanent commission, it might be better to set up discussion groups on the issues involved. He was, of course, completely correct in his belief that it would be much easier to convene many short conferences than to found a standing body, but such meetings could hardly be expected to have a significant worldwide or national impact; they are unlikely to be able to generate much public debate or authoritatively advance policy alternatives.

The mass media were not crying out for regulation in this area, Congress was not enacting an American commission, and the international organizations were passive, afraid to initiate a worldwide body. The assurance of financial backing, I felt, could make a difference; if funds could be raised—from a foundation, perhaps—to finance a commission, the international organizations strapped for monies might suddenly discover a way to work with a health-ethics commission (although even then, no doubt, they would try to keep it at arm's length).

I sent letters to several major foundations and to several lesser ones, asking for their support. Some answered that their interests lay elsewhere. Still others pointed out that they were already supporting the work of individual scholars in this area. The Ford and Rockefeller foundations wanted to explore the matter further, but during the course of these discussions, it became clear that they might support a few "workshops" but not a permanent national or worldwide commission on health ethics.

Though I still felt that such workshops were no substitute for the needed commission, I did realize the value of bringing together groups of researchers, practitioners, officials, theologians, and humanists who could work out specific guidelines for specific fields (e.g., the use of amniocentesis). Such workshops could serve as stopgaps until commissions *continuously at work and publicly instituted* could be set up. And through their very endeavors, a wider recognition of the need for ongoing, authoritative, public deliberations might evolve. As I write these lines, however, no foundation has yet made a final decision to support such workshops.

Meanwhile, the 92nd session of Congress ended on October 18, 1972, before the House Committee got a chance to review the bill and pass it on to the floor. This spelled an automatic, albeit possibly temporary, death of the Mondale bill, which called for a two-year Congressional health-ethic study commission. By congressional procedures, if the Senate acts but the House does not, the bill dies as the session ends—even if, as in this case, it was unanimously endorsed by the Senate. The bill, however, can be resurrected in the next session of Congress if the Senate endorses it again.

The lack of interest on the part of the media, international organizations, foundations, and the House of Representatives in dealing with the Mondale bill was rather discouraging. I briefly considered focusing my public efforts on one of the other policy issues close to my heart. Then I came upon several fine examples of the good health-ethic commissions could do, and these reinforced my conviction that the tools of more effective and responsive societal guidance were sorely needed, however negligible public awareness or political support was at present. These forerunners of the commission are limited in scope, support, funds, and staff; their contributions thus illustrate both the need for a full-fledged, representative, adequately endowed and adequately staffed commission, and the virtue of carrying out more limited activities until such a body will be established.

The Hastings Report on Mass Screening

Following the publication of the report in *Science*, Peter Steinfels, who used to work for the Catholic intellectual journal, *Commonweal*, called to learn more about the Paris meeting. He is now with the Institute of Society, Ethic and Life Sciences, often referred to as the Hastings Institute (it is located in Hastings, New York). Steinfels, who is young, bearded, and ascetic-looking, turned up with another young person, no tie, uncurbed curls, the Director of the Hastings Institute, Daniel Callahan, also a Catholic intellectual.

After we discussed the Paris meeting and the numerous essays that the Hastings Institute's fellows had prepared in the area of health ethics, the discussion turned to a project which had already yielded the desired fruit: policy guidelines. The institute had issued a report on the ethical and social issues raised by screening large numbers of people for genetic disease (see Appendix 8). Callahan, who did most of the talking, explained that "the group who formulated the guidelines for mass screening was mostly opposed to the whole idea but favored a cautious and careful approach."

The report provides a set of criteria for assessing the merits of a particular screening program, criteria which all seem very sensible and self-evident, but which were nevertheless often ignored by those who set up mass-screening programs before the report was available, and which—for reasons I'll return to—since its publication, continue to be overlooked.

The report's most basic criterion is that no program be set up before adequate testing procedures are available "to avoid the problems that occurred initially in PKU screening." (In these tests, many children were wrongly identified as having the disease, and quite a few who did have it were diagnosed healthy.)

The sickle-cell testing program, recently introduced in a great hurry, without a thorough review by any board, fails by the Hastings criteria. The programs test either school children, at an age when the illness very often has already struck, or newborns, a stage at which detection is difficult. Tests of couples considering having a child would make much more sense, although such programs are more difficult to administer than school programs in which the children can be lined up at will.

Besides asking for safe tests, the Hastings group also called attention to *a risk of possible psychological or social injury*. The question is: How harmful will the "labeling" of persons be? As the result of mass-screening tests people will be labeled as carriers of sick genes, which may harm their social standing and their view of themselves. Social scientists vary on the degree of importance they attach to the ways people come to view themselves and the ways they are viewed by others (for instance, who is branded a "criminal" and who a "law-abiding" citizen?). However, social science data leaves no doubt that at least in some areas, labeling has rather serious consequences.

An often-cited example of the effects of labeling is a study by Dr. Robert Rosenthal undertaken at Harvard. He asked a class of psychologists to run rats through a maze. While all the rats were of the same ordinary "garden variety," the class was told that half of them were a special quick-to-learn breed. Those who were assigned the "brighter" breed were reported to have exhibited a much friendlier and gentler attitude toward their animals than the part of the class that dealt with those labeled "slow learners." Moreover, the rats branded "bright" got a mean number of 2.32 correct responses in the maze, while the "dull" ones got only 1.54.[1] When the experiment was applied to teachers in a school, who were told that some children were found to be particularly prone to improvement, these pupils did better than the comparable control group by 3.8 points in a weighted means.[2] Other studies show similar results.

In regard to the issue at hand, there is little doubt that if young kids are told that they have an XYY chromosome structure (which occurs in about one of every thousand males born), and which has been repeatedly reported as being associated with a predisposition toward serious deviant behavior,[3] they could easily begin to view themselves as having a criminal destiny. Moreover, if parents are told that their child carries the XYY chromosome, many may well come to suspect normal assertive moves of the children as manifestations of their criminal potential; consequently, they may push their kids–whatever the influence of their genes–into an aggressive, ultimately criminal, personality and way of life.

Beyond parents, teachers, and self-image, such labeling is likely to affect the attitudes of practically everyone who knows about a person's genetic test scores. This is no longer a hypothetical consideration. The undesirable consequences, which the Hastings group warned were possible, have already made themselves felt. In 1971 the state of Massachusetts, responding to the demands of black community leaders and their white supporters, passed a law requiring that all school-age children be tested for the sickle-cell trait, a trait relatively common among black children (it hits one out of every five hundred)[4] and very rare among other groups. A dozen states rushed to follow suit. The trait is harmless by itself, but when both parents have it, there is a one in four chance that their child will have the horrible disease (it first causes pain, then deterioration of major organs such as the liver and kidneys).[5]

The results of the tests have been used to brand carriers of the disease as poor employment risks[6] and poor risks for life insurance.[7] Recently many black people started wondering whether or not the undesirable consequences of the test outweighed its benefits, especially as there is, so far, no known cure for the illness.

Related questions must be asked about other genetic tests that are being increasingly used, promoted, or sought. Screening programs for carriers of the gene for Tay-Sachs disease was started among Jewish people in Baltimore in 1971. A screening test for Cooley's anemia, relatively more common among people of Mediterranean descent, is being developed. A new genetic test is about to be marketed which will screen people's susceptibility to emphysema, a degenerative lung disease.[8] Tests for dysautonomia (a disease which chiefly affects Ashkenazi Jews) and cystic fibrosis (which hits one out of every thousand Caucasian babies born in the United States),[9] are being actively sought. Dozens of other tests are likely to follow. If they are to do more good than harm, there must be a mechanism for reviewing the programs before they are enacted. The

Hastings criteria, formulated by a private group, do not command the support and power that the criteria formulated by a public authority could provide.

The Hastings report also points out that it is necessary to accompany such a new program with *carefully designed and executed public information programs*. Experience shows that the public—and even some doctors—confuse the sicke-cell *trait* [10]—again, quite harmless by itself—with the disease of sickle-cell *anemia*, which is found only when *both* parents have the trait and then only in approximately one out of four of their offspring.

The Hastings report also suggests other criteria for evaluating or designing mass genetic screening tests; these include equal access, absence of compulsion, and informed consent. (For additional details, see Appendix 8.) Again, the record shows that these recommendations are not often followed.

This particular Hastings report (other ones are being formulated) is not all-comprehensive. For instance, it does not deal with the question, How safe is safe?—an essential issue for new tests—or, How can safety be tested before tests are used en masse? Secondly, because the report is based chiefly on deliberations and dialogue, it shows little benefit from research and empirical input to back up its suggestions. Nevertheless it is of immense value, if only because it provides all those who will listen with a detailed list of what must be taken into account before such programs are initiated.

The main source of the weakness of the Hastings Institute's efforts is not in the Hastings group itself, but in the absence of an authority in Washington. No private group can possibly have the necessary national visibility and clout. If similar efforts were undertaken by a national commission composed of leading authorities in the respective fields and representative members of the community backed up by congressional status and a staff, they would command a much greater following. Of course, even if such a national body were formed, private groups would still have to continue their deliberations. These issues must be as widely discussed as possible, for a continuous dialogue of many divergent viewpoints is essential if the bases for a new ethic as well as for policy guidelines are to evolve. A national body would provide a much-needed focus for such private deliberations, but it could not, nor should it try, to replace them.

The Patient's Bill of Rights

Another development illustrates how, lacking the benefit of an institutionalized review mechanism, the nation tried, with partial success, to cope with its need to review and form policy in the health and genetic fields. The American Hospital Association issued the bill of rights for patients, first in November 1972 and again in January 1973, to its seven thousand member hospitals. The bill was formulated by a committee appointed by the trustees of the American Hospital Association and discussed by its regional advisory boards, composed of hospital administrators. Consumer representatives were also involved in the committee's work. The bill's twelve-point protocol, given in full in Appendix 3, is summarized as follows:

"1. The patient has the right to considerate and respectful care.

2. The patient has the right to obtain from his physician complete current information concerning his diagnosis, treatment, and prognosis in terms the patient can reasonably be expected to understand.

3. The patient has the right to receive from his physician information necessary to give informed consent prior to the start of any procedure and/or treatment.

4. The patient has the right to refuse treatment to the extent permitted by law, and to be informed of the medical consequences of his action.

5. The patient has the right to every consideration of his privacy concerning his own medical care program.

6. The patient has the right to expect that all communications and records pertaining to his care should be treated as confidential.

7. The patient has the right to expect that within its capacity a hospital must make reasonable response to the request of a patient for services.

8. The patient has the right to obtain information as to any relationship of his hospital to other health care and educational institutions insofar as his care is concerned.

9. The patient has the right to be advised if the hospital proposes to engage in or perform human experimentation affecting his care or treatment.

10. The patient has the right to expect reasonable continuity of care.

11. The patient has the right to examine and receive an explanation of his bill, regardless of source of payment.

12. The patient has the right to know what hospital rules and regulations apply to his conduct as a patient."

Several hospitals adopted the bill, and at least two (Boston's Beth Israel Hospital and New York's Martin Luther King Health Center) now provide their patients with their own version of it; but most hospitals did not embrace it. And yet, the charter is certainly a valuable one. Both technological and social developments have rendered the existing hospital structure virtually obsolescent, and there is a particularly great need for a new definition of the relationship between the patient and the institution.

The fact that those who administer hospitals took the initiative in preparing this bill is hardly surprising, since there is no community-based body to assume such duties. However, it must also be noted that the charter, deprived of public hearings of the kind a congressional committee would have generated, was not subject to wide discussion or public involvement; it is no wonder, then, that it is easy to ignore. Moreover, the fact that it was formulated by a board composed chiefly of those in power will hardly reassure the more activist "consumer" groups. (Actually, Dr. Willard Gaylin went so far as to indicate that the document "perpetuates the very paternalism that precipitated the abuses.") [11] A more widely representative body would have given the bill more authority.

Like most documents formulated chiefly to express a sentiment and to affirm a position ("We should pay more attention to patients' rights"), the charter is rather long on general statements and somewhat short on attention to specifics, and yet it is the latter that are essential if it is to be widely used. For example, the statement, "The patient has the right to refuse treatment. . . ." is qualified by the phrase "to the extent permitted by law," as though the law provided a clear guideline. Actually, if the patients themselves have the right to insist, for instance, that life-extending machines be turned off, they must be conscious when they so choose; but then their action would be tantamount to suicide. On the other hand, if the patient has to be unconscious beyond recall before the machines can be turned off, the right to refuse service is not his. Who, then, exercises the right? One doctor? Two? Three? With, or without, consultation of the next of kin? Under what medical conditions?

The Medical Society of the State of New York suggested adding the clause "irrefutable evidence that biological death is inevitable," [12] but such evidence may come long before a person loses consciousness. The society also suggested adding the phrase "is the decision of the patient and/or the immediate family with the approval of the family physician."[13] But what if there is no family physician? And should not at least one other doctor, not as deeply involved, be consulted? Clearly the Patient's Bill of Rights leaves these and many other issues unresolved.

If the authors of the bill had expressed greater concern with mechanisms of implementation, they would have been more aware of the need for local health-ethics boards to review decisions made to "turn off" lives; the need for national and international boards to formulate guidelines; and a research staff to study the actual results of various steps undertaken, in order to apply them in future deliberations. (There are review committees inside hospitals but these, with few exceptions, are limited to physicians only and only to those of that hospital. This insularity tends to limit their critical power.)

This example will have to stand for scores of others, all resembling the Patient's Bill of Rights, as a document which, though well-intentioned and encouraging, has not been sufficiently "processed" or provided with the mechanisms (e.g., local review boards) for its thorough implementation.

The Abortion Ruling

Another example of the need in the area of health-ethics for a better backup by national decisions occurred on January 22, 1973, when the U.S. Supreme Court overruled all state laws that prohibit or restrict a woman's right to obtain an abortion during her first three months of pregnancy. It is now up to the woman and her physician to decide what course to follow. For the last six months of pregnancy, abortion can be "regulated" by the states to secure maternal health (e.g., they can limit abortions to qualified facilities). Only in the last ten weeks of pregnancy, when the fetus is judged capable of surviving if born ("viable"), may a state prohibit abortion. In effect, the court overruled the laws which severely restricted abortions in most of the states of the Union.

This welcome act also turned a matter that was previously controlled by the government over to individual choice. It also left it up to agencies other than governmental ones to worry about, and to inform parents, of the risks involved in abortions, which are now estimated to be undergone by 1.6 million American women each year. [14] These risks are not trivial. While an abortion performed by a well-trained physician during the first twelve weeks of pregnancy is said to be safer than a tonsillectomy or an actual birth (out of 100,000 patients, the death rate for abortion is 2; for tonsillectomies, 17; for the pregnancy, delivery, and post-natal periods, 20),[15] in an abortion done in the second trimester,[16] complications are three to four times more likely to arise, according to one source.[17]

It is not the Supreme Court's business, following such a ruling as its one on abortion, to arrange the necessary public education campaign—for instance, advising those who either use no contraceptives or rather unreliable means not to rely on abortion for birth control, or cautioning those who need an abortion not to put it off. This, it might be said, is the job of HEW. But a public authority could go a long way to see to it that matters the Court leaves undone will indeed be picked up by the appropriate executive agency, and with the desired vigor and scope.

Next the ruling involves an empirical matter. By getting the state out of the business of regulating abortions in the first twelve weeks of pregnancy, the Supreme Court, in effect, also allows *any* M.D., not just a gynecologist, to perform an abortion. Several leading doctors, including Dr. Morton A. Schiffer and Dr. Bernard Nathanson,[18] implied that it would have been better to limit the practice of abortion to (a) qualified doctors, and (b) hospitals and clinics appropriately equipped and staffed.

The issue here is not whether they are right or the Court is, but that there is no systematic procedure through which the relevant medical data and considerations on this or any other matter are regularly brought before the Court before it rules. The Court will tolerate experts as "friends of the court," but this procedure is occasional rather than systematic; further, it tends to attract individuals, voluntary associations, or civic groups, and only very rarely institutional "think tanks," with their data banks and research staffs. The Court, in these matters, is simply obsolescent; like Congress, it follows the same procedures used a half century ago, before the knowledge explosion (when one or a few experts knew more than you needed to know about a particular area), and before computers, with their memories and analysts.

Medical Review Boards

Another highly relevant development came from a very different direction. In October 1972, Congress enacted a bill widely referred to as "H.R. 1," a large package of amendments to the Social Security Act. The numerous clauses of the bill run into 940 pages, and among these is Section 249F, barely known to the general public. The amendment calls for setting up "professional standards review organizations" (PSRO's). The basic idea is to subject hospitals and other health units to outside review of not only the proper use of funds—the typical accountability expected and required of anyone who uses public monies—but also of professional, that is, medical, matters. The main motive seems to be to reduce the number of the poor and the aged who are sent to hospitals by doctors and whose care is charged to the taxpayers. (The amendment calls for checking nonemergency cases with the PSRO's before admission.) At the same time, the provision opens the door, in principle, to outside, or perhaps even public, scrutiny of what doctors are doing.

The law is rather vague as to who is to provide these outside review boards. But the basic assumption is that doctors will oversee doctors. Even this is quite innovative, because many doctors feel they need no overview and that if review is to take place, it should be by their peers—that is, by people who are equal to them in status, who are usually members of the same hospital staff or local medical society, and who are often—like those of the peer age—rather beholden to each other. Actually, peer reviews are often surprisingly strict. But don't we need more?

The PSRO's, which are to be established throughout the United States by January 1, 1976, go one step further by calling for "outsiders" to review insiders and to act as a kind of medical audit. But if these outsiders are chosen by the local medical societies, they may not be as independent as they should be of those they are to review.

Above all, it seems desirable that the PSRO's should include not only doctors but also community representatives and "specialists" in societal and ethical matters to make sure that "consumer," social, and moral issues be taken into account and to counteract any self-serving tendencies of the doctors.

The PSRO's represent an important spot at which to enter the closed professional system, because unlike community advisory boards set up around hospitals or comprehensive health planning agencies—both of which are quite welcome—the PSRO's will be able to control the main flow of taxpayers' funds to health units and hence carry much more weight.

The Mondale Bill—Revived and Beyond

As the 93rd session of Congress got underway in early 1973, Senator Mondale reintroduced his bill, now numbered "Senate Resolution 71." The Senate seemed likely to approve it again, but no one could make any predictions as to what the House would do.

Also early in 1973, Sen. Edward Kennedy held extensive hearings on an issue that a health-ethics commission, had it existed, would have dealt with: the conditions under which experimentations with human subjects can be tolerated. The press was again filled with gory reports about this or that ethical violation, but paid little attention to the more general questions concerning how any regulations could be implemented.

But progress was made during the hearing—two colleagues called for the establishment of a more advanced and more potent health-ethics commission than the Mondale bill outlined. Dr. Bernard Barber testified that he favored ". . . the establishment of a National Board of Biomedical Research Ethics. As members of that board I would like to see not only members of the medical research profession, who are of course indispensable, but also people who are outsiders to the profession and who represent the public. These outsiders cannot be ordinary men-in-the-street or men given to absolute morals; they should be *informed outsiders*—lawyers or social scientists who have the expertise to deal with the fact that medical research ethics are also social and not just medical matters. . . . The board could define goals, establish institutions and mechanisms, and provide necessary monitoring for standards and practices that are only what the profession rightly values and the public increasingly and rightly demands."[19]

Dr. Jay Katz, adjunct Professor of Law and Psychiatry at Yale, suggested that a permanent body be established to regulate all federally supported research involving human subjects. Such a board, Katz said, should be independent from the government, since much experimentation that requires supervision is carried out in government-owned laboratories. He wanted the President to appoint the board, and suggested that "its members should come from many disciplines, including representatives from the public at large," and that the board should have "regulatory authority," that is, it should formulate policy and set up the necessary regulations and mechanisms to promote them.[20]

Note that the concern in the Kennedy hearings focused on those relatively few persons who are subjects in experiments. My feeling is that we are all "subjects"—the millions who take the Pill; the millions who do not receive genetic counseling, the millions exposed to food additives which may well be cancer-inducing, etc., etc. We need to develop a more effective review mechanism of all illness-producing and illness-preventing forces in our life. The focus on human subjects in laboratories should be the opening wedge, not a conciliatory gesture that gives reprieve from much-needed nationwide, not just lab-wide, scrutiny.

It is up to us all to see that the reforms will not stop here. The efforts to form effective and responsive overview mechanisms cannot be advanced by a few senators and professors. Their future depends on citizens being informed and alerted to the regulatory functions—beyond the well-popularized "human interest" stories —and on the general public, led by active groups of citizens, taking on this issue, as they previously took on those for peace in Vietnam, civil rights, and pollution control.

Action is needed on several fronts. On the national level, Congress must be urged to set up a permanent National Health-Ethics Commission which will include members of a variety of disciplines, not just medicine, and representatives of the public, and which will be backed up by a research staff.

Locally, each state, city, and town needs a local review Health-Ethics Board to oversee its hospitals and clinics, its medical healers and researchers.

Individually, citizens and their leaders have to become better informed about new medical and genetic developments and the issues raised by their effects on matters of illness and health, life and death. The people must be the guards of the guards. Professionals cannot be left to be guided only by their own lights and those of their peers.

Aside from carrying out their citizens' duty, individuals should inform themselves more on these matters for their own protection and for the protection of those dear to them. Otherwise, they will surely not reap the full benefits of the new developments nor be spared the many dangers.

Ultimately, only when citizens learn more about the ways society may be directed to respond to their needs, and only when they act armed with this new knowledge rather than following the interest or preconceptions of the few, will the country be managed for their well-being. Casting a vote once every few years is simply no longer enough to secure a government responsive to the people and truly concerned with our future. An informed and active citizenry, dealing with national and local governments as well as the administrative boards of health institutions has become a prerequisite not just for a sound democracy but for a healthy body, a normal child, and indeed, for life itself.

The FDA Warning on Birth Control Pills

(Full original text.)

What You Should Know About Birth Control Pills (Oral Contraceptive Products)

All of the oral contraceptive pills are highly effective for preventing pregnancy, when taken according to the approved directions. Your doctor has taken your medical history and has given you a careful physical examination. He has discussed with you the risks of oral contraceptives, and has decided that you can take this drug safely.

This leaflet is your reminder of what your doctor has told you. Keep it handy and talk to him if you think you are experiencing any of the conditions you find described.

A Warning About "Blood Clots"

There is a definite association between blood-clotting disorders and the use of oral contraceptives. The risk of this complication is six times higher for users than for non-users. The majority of blood-clotting disorders are not fatal. The estimated death rate from blood-clotting in women *not* taking the pill is one in 200,000 each year; for users, the death rate is about six in 200,000. Women who have or who have had blood clots in the legs, lung, or brain should not take this drug. You should stop taking it and call your doctor immediately if you develop severe leg or chest pain, if you cough up blood, if you experience sudden and severe headaches, or if you cannot see clearly.

Who Should Not Take Birth Control Pills

Besides women who have or who have had blood clots, other women who should not use oral contraceptives are those who have serious liver disease, cancer of the breast, or certain other cancers, and vaginal bleeding of unknown cause.

Special Problems

If you have heart or kidney disease, asthma, high blood pressure, diabetes, epilepsy, fibroids of the uterus, migraine headaches, or if you have had any problems with mental depression, your doctor has indicated you need special supervision while taking oral contraceptives. Even if you don't have special problems, he will want to see you regularly to check your blood pressure, examine your breasts, and make certain other tests.

When you take the pill as directed, you should have your period each month. If you miss a period, and if you are sure you have been taking the pill as directed, continue your schedule. If you have not been taking the pill as directed and if you miss one period, stop taking it and call your doctor. If you miss two periods see your doctor even though you have been taking the pill as directed. When you stop taking the pill, your periods may be irregular for some time. During this time you may have trouble becoming pregnant.

If you have had a baby which you are breast feeding, you should know that if you start taking the pill its hormones are in your milk. The pill may also cause a decrease in your milk flow. After you have had a baby, check with your doctor before starting to take oral contraceptives again.

What to Expect

Oral contraceptives normally produce certain reactions which are more frequent the first few weeks after you start taking them. You may notice unexpected bleeding or spotting and experience changes in your period. Your breasts may feel tender, look larger, and discharge slightly. Some women gain weight while others lose it. You may also have episodes of nausea and vomiting. You may notice a darkening of the skin in certain areas.

Other Reactions to Oral Contraceptives

In addition to blood clots, other reactions produced by the pill may be serious. These include mental depression, swelling, skin rash, jaundice or yellow pigment in your eyes, increase in blood pressure, and increase in the sugar content of your blood similar to that seen in diabetes.

Possible Reactions

Women taking the pill have reported headaches, nervousness, dizziness, fatigue, and backache. Changes in appetite and sex drive, pain when urinating, growth of more body hair, loss of scalp hair, and nervousness and irritability before the period also have been reported. These reactions may or may not be directly related to the pill.

Note About Cancer

Scientists know the hormones in the pill (estrogen and progesterone) have caused cancer in animals, but they have no proof that the pill causes cancer in humans. Because your doctor knows this, he will want to examine you regularly.

Remember

While you are taking ——, call your doctor promptly if you notice any unusual change in your health. Have regular checkups and your doctor's approval for a new prescription.

Dr. Virginia Apgar's Guidelines for Prospective Parents*

1. An individual, or couple, who thinks that a close relative has a disorder which might be hereditary should take advantage of genetic counseling.

2. The ideal age for a woman to have children is between 20 and 35. If possible, it is best not to begin having babies before the age of 18 and to complete childbearing before age 40.

3. A man should beget his children before he reaches the age of 45.

4. There should be an interval of at least two years between the end of one pregnancy and the beginning of another.

5. With every subsequent child, beginning with the third, there is increasing hazard of stillbirth, congenital malformation and prematurity.

6. When a couple plans to conceive a child, intercourse should take place at intervals of no more than 24 hours for several days just preceding and during the estimated time of ovulation.

7. Every pregnant woman needs good prenatal care supervised by a physician who keeps current on new medical research in tera-

*These guidelines, together with a more detailed discussion can be found in Virginia Apgar and Joan Beck, *Is My Baby All Right?* (New York: Trident Press, 1972), pp. 435–452.

Copyright ☻ 1972 by Joan Beck. Reprinted by permission of Trident Press, a division of Simon & Shuster, Inc. Dr. Apgar is Clinical Professor of Pediatrics at Cornell University Medical College and Vice-President for Medical Affairs of the National Foundation-March of Dimes.

tology and fetology and who will help her deliver her baby in a reputable, up-to-date hospital.

8. No woman should become pregnant unless she is sure she has had rubella or has been effectively immunized against it.

9. From the very beginning of pregnancy, a woman should do everything possible to keep herself in good health and to avoid exposure to contagious diseases.

10. All during pregnancy, a woman should avoid eating under-cooked red meat or contact with any cat which might be the source of a toxoplasmosis infection.

11. A woman who is pregnant, or thinks she could possibly be pregnant, should not take any drugs whatsoever unless absolutely essential—and then only when prescribed by a physician who is aware of the pregnancy.

12. An X-ray examination or radiation treatment should not be given to any pregnant woman or to any woman who thinks there is the slightest possibility she might be pregnant.

13. Cigarettes should not be smoked during pregnancy.

14. A nourishing diet, rich in proteins, vitamins, and minerals and adequate in total calories, is essential during pregnancy.'

15. A prospective mother who is Rh negative should make sure her physician takes the necessary steps to protect her unborn baby and subsequent children from Rh disease.

16. Every precaution should be taken to prevent a baby from being born prematurely.

17. Good obstetrical care in a well-equipped hospital can greatly reduce the hazards of being born.

APPENDIX 3

A Patient's Bill of Rights*

**(Affirmed by the Board of Trustees of the
American Hospital Association November 17, 1972.
Approved by the American Hospital Association's
House of Delegates on February 6, 1973.)**

The American Hospital Association presents a Patient's Bill of Rights with the expectation that observance of these rights will contribute to more effective patient care and greater satisfaction for the patient, his physician, and the hospital organization. Further, the Association presents these rights in the expectation that they will be supported by the hospital on behalf of its patients, as an integral part of the healing process. It is recognized that a personal relationship between the physician and the patient is essential for the provision of proper medical care. The traditional physician-patient relationship takes on a new dimension when care is rendered within an organizational structure. Legal precedent has established that the institution itself also has a responsibility to the patient. It is in recognition of these factors that these rights are affirmed.

1. The patient has the right to considerate and respectful care.

2. The patient has the right to obtain from his physician complete current information concerning his diagnosis, treatment, and prognosis in terms the patient can be reasonably expected to understand. When it is not medically advisable to give such information to the patient, the information should be made available to an appropriate person in his behalf. He has the right to know by name, the physician responsible for coordinating his care.

3. The patient has the right to receive from his physician infor-

*Reprinted by permission of the American Hospital Association.

mation necessary to give informed consent prior to the start of any procedure and/or treatment. Except in emergencies, such information for informed consent, should include but not necessarily be limited to the specific procedure and/or treatment, the medically significant risks involved, and the probable duration of incapacitation. Where medically significant alternatives for care or treatment exist, or when the patient requests information concerning medical alternatives, the patient has the right to such information. The patient also has the right to know the name of the person responsible for the procedures and/or treatment.

4. The patient has the right to refuse treatment to the extent permitted by law, and to be informed of the medical consequences of his action.

5. The patient has the right to every consideration of his privacy concerning his own medical care program. Case discussion, consultation, examination, and treatment are confidential and should be conducted discreetly. Those not directly involved in his care must have the permission of the patient to be present.

6. The patient has the right to expect that all communications and records pertaining to his care should be treated as confidential.

7. The patient has the right to expect that within its capacity a hospital must make reasonable response to the request of a patient for services. The hospital must provide evaluation, service, and/or referral as indicated by the urgency of the case. When medically permissible a patient may be transferred to another facility only after he has received complete information and explanation concerning the needs for and alternatives to such a transfer. The institution to which the patient is to be transferred must first have accepted the patient for transfer.

8. The patient has the right to obtain information as to any relationship of his hospital to other health care and educational institutions insofar as his care is concerned. The patient has the right to obtain information as to the existence of any professional relationships among individuals, by name, who are treating him.

9. The patient has the right to be advised if the hospital proposes to engage in or perform human experimentation affecting his care or treatment. The patient has the right to refuse to participate in such research projects.

10. The patient has the right to expect reasonable continuity of

care. He has the right to know in advance what appointment times and physicians are available and where. The patient has the right to expect that the hospital will provide a mechanism whereby he is informed by his physician or a delegate of the physician of the patient's continuing health care requirements following discharge.

11. The patient has the right to examine and receive an explanation of his bill regardless of source of payment.

12. The patient has the right to know what hospital rules and regulations apply to his conduct as a patient.

No catalogue of rights can guarantee for the patient the kind of treatment he has a right to expect. A hospital has many functions to perform, including the prevention and treatment of disease, the education of both health professionals and patients, and the conduct of clinical research. All these activities must be conducted with an overriding concern for the patient, and, above all, the recognition of his dignity as a human being. Success in achieving this recognition assures success in the defense of the rights of the patient.

A Statement on Death (Declaration of Sydney)*

(Adopted by the 22nd World Medical Assembly, Sydney, Australia, August 1968.)

1. The determination of the time of death is in most countries the legal responsibility of the physician and should remain so. Usually he will be able without special assistance to decide that a person is dead, employing the classical criteria known to all physicians.

2. Two modern practices in medicine, however, have made it necessary to study the question of the time of death further: the ability to maintain by artificial means the circulation of oxygenated blood through tissues of the body which may have been irreversibly injured and the use of cadaver organs such as heart or kidneys for transplantation.

3. A complication is that death is a gradual process at the cellular level with tissues varying in their ability to withstand deprivation of oxygen. But clinical interest lies not in the state of preservation of isolated cells but in the fate of a person. Here the point of death *of the different cells and organs* is not so important as the certainty that the process has become irreversible by whatever techniques of resuscitation that may be employed.

4. This determination will be based on clinical judgment supplemented *if necessary* by a number of diagnostic aids of which the electroencephalograph is currently the most helpful. However, no single technological criterion is entirely satisfactory in the present state of medicine nor can any one technological procedure

*Reprinted by permission of The World Medical Association, Inc.

be substituted for the overall judgment of the physician. *If transplantation of an organ is involved, the decision that death exists should be made by two or more physicians and the physicians determining the moment of death should in no way be immediately concerned with the performance of the transplantation.*

5. Determination of the point of death of the person makes it ethically permissible to cease attempts at resuscitation, and in countries where the law permits, to remove organs from the cadaver provided that prevailing legal requirements of consent have been fulfilled.

Recommendations Guiding Doctors in Clinical Research (Declaration of Helsinki) *

(Adopted by the 18th World Medical Assembly, Helsinki Finland, 1964)

Introduction

It is the mission of the doctor to safeguard the health of the people. His knowledge and conscience are dedicated to the fulfillment of this mission.

The Declaration of Geneva of the World Medical Association binds the doctor with the words: "The health of my patient will be my first consideration" and the International Code of Medical Ethics which declares that "Any act or advice which could weaken physical or mental resistance of a human being may be used only in his interest."

Because it is essential that the results of laboratory experiments be applied to human beings to further scientific knowledge and to help suffering humanity, The World Medical Association has prepared the following recommendations as a guide to each doctor in clinical research. It must be stressed that the standards as drafted are only a guide to physicians all over the world. Doctors are not relieved from criminal, civil and ethical responsibilities under the laws of their own countries.

In the field of clinical research a fundamental distinction must be recognized between clinical research in which the aim is essentially

*Reprinted by permission of The World Medical Association, Inc.

therapeutic for a patient, and the clinical research, the essential object of which is purely scientific and without therapeutic value to the person subjected to the research.

I. Basic Principles

1. Clinical research must conform to the moral and scientific principles that justify medical research and should be based on laboratory and animal experiments or other scientifically established facts.

2. Clinical research should be conducted only by scientifically qualified persons and under the supervision of a qualified medical man.

3. Clinical research cannot legitimately be carried out unless the importance of the objective is in proportion to the inherent risk to the subject.

4. Every clinical research project should be preceded by careful assessment of inherent risks in comparison to foreseeable benefits to the subject or to others.

5. Special caution should be exercised by the doctor in performing clinical research in which the personality of the subject is liable to be altered by drugs or experimental procedure.

II. Clinical Research Combined with Professional Care

1. In the treatment of the sick person, the doctor must be free to use a new therapeutic measure, if in his judgment it offers hope of saving life, reestablishing health, or alleviating suffering.

If at all possible, consistent with patient psychology, the doctor should obtain the patient's freely given consent after the patient has been given a full explanation. In case of legal incapacity, consent should also be procured from the legal guardian; in case of physical incapacity the permission of the legal guardian replaces that of the patient.

2. The doctor can combine clinical research with professional care, the objective being the acquisition of new medical knowledge,

only to the extent that clinical research is justified by its therapeutic value for the patient.

III. Non-Therapeutic Clinical Research

1. In the purely scientific application of clinical research carried out on a human being, it is the duty of the doctor to remain the protector of the life and health of that person on whom clinical research is being carried out.

2. The nature, the purpose and the risk of clinical research must be explained to the subject by the doctor.

3a. Clinical research on a human being cannot be undertaken without his free consent after he has been informed; if he is legally incompetent, the consent of the legal guardian should be procured.

3b. The subject of clinical research should be in such a mental, physical and legal state as to be able to exercise fully his power of choice.

3c. Consent should, as a rule, be obtained in writing. However, the responsibility for clinical research always remains with the research worker; it never falls on the subject even after consent is obtained.

4a. The investigator must respect the right of each individual to safeguard his personal integrity, especially if the subject is in a dependent relationship to the investigator.

4b. At any time during the course of clinical research the subject or his guardian should be free to withdraw permission for research to be continued.

The investigator or the investigating team should discontinue the research if in his or their judgment, it may, if continued, be harmful to the individual.

Author's note: A much more detailed statement on these issues is available from the U.S. Department of Health, Education and Welfare: "The Institutional Guide to DHEW Policy on Protection of Human Subjects," (Washington, D.C.: U.S. Government Printing Office, 1971). See also Jay Katz, ed., *Experimentation with Human Beings* (New York: Russell Sage Foundation, 1972).

A National Advisory Commission on Health, Science and Society (The Mondale Bill)

(The draft of a bill [S.J. 75] to set up a Congressional commission, introduced to the 92nd Congress by Sen. Walter Mondale [Minnesota] and a group of other senators, which was unanimously approved by the Senate but not by the House. The bill has been reintroduced in the 93rd Congress [S.J. Res. 71].)

The Committee on Labor and Public Welfare, to which was referred the resolution (S.J. Res. 75) to provide for a study and evaluation of the ethical, social, and legal implications of advances in biomedical research and technology, having considered the same, reports favorably thereon with an amendment in the nature of a substitute and recommends that the resolution as amended do pass.

Committee Amendment

The amendment is as follows:

That this joint resolution may be cited as the "National Advisory Commission on Health Science and Society Resolution".

ESTABLISHMENT OF COMMISSION

Sec. 2. There is hereby established a National Advisory Commission on Health Science and Society (hereinafter referred to as the "Commission").

MEMBERSHIP

Sec. 3. (a) The Commission shall be composed of fifteen members to be appointed by the President from the general public and from individuals in the fields of medicine, law, theology, biological science, physical science, social science, philosophy, humanities, health administration, government, and public affairs.

(b) Any vacancy in the Commission shall not affect its powers.

(c) The President shall designate one of the members to serve as Chairman and one to serve as Vice Chairman of the Commission.

(d) Eight members of the Commission shall constitute a quorum.

DUTIES OF THE COMMISSION

Sec. 4. (a) The Commission shall undertake a comprehensive investigation and study of the ethical, social, and legal implications of advances in biomedical research and technology, which shall include, without being limited to—

(1) analysis and evaluation of scientific and technological advances in the biomedical sciences, past, current and projected;

(2) analysis and evaluation of the implications of such advances, both for individuals and for society;

(3) analysis and evaluation of laws, codes, and principles governing the use of technology in medical practice;

(4) analysis and evaluation through the use of seminars and public hearings and other appropriate means, of public understanding of attitudes toward such implications; and

(5) analysis and evaluation of implications for public policy of such findings as are made by the Commission with respect to biomedical advances and public attitudes toward such advances.

(b) The Commission shall make maximum feasible use of related investigations and studies conducted by public and private agencies.

(c) The Commission shall transmit to the President and to the Congress one or more interim reports and, not later than two years after the first meeting of the Commission, one final report, containing detailed statements of the findings and conclusions of the Commission, together with its recommendations, including such recommendations for action by public and private bodies and individuals as it deems advisable.

POWERS OF THE COMMISSION

Sec. 5. (a) The Commission or, on the authorization of the Commission, any subcommittee or members thereof, may, for the purpose of carrying out the provisions of this joint resolution, hold such hearings, take such testimony, and sit and act at such times and places as the Commission deems advisable. Any member authorized by the Commission may administer oaths or affirmations to witnesses appearing before the Commission or any subcommittee or members thereof.

(b) Each department, agency, and instrumentality of the executive branch of the Government, including independent agencies, is authorized and directed, to the extent permitted by law, to furnish to the Commission, upon request made by the Chairman or Vice Chairman, such information as the Commission deems necessary to carry out its functions under this joint resolution.

(c) Subject to such rules and regulations as may be adopted by the Commission, the Chairman shall have the power to—

(1) appoint and fix the compensation of an executive director, and such additional staff personnel as he deems necessary, without regard to the provisions of title 5, United States Code, governing appointments in the competitive service, and without regard to the provisions of chapter 51 and subchapter III of chapter 53 of such title relating to classification and General Schedule pay rates, but at rates not in excess of the maximum rate for GS-18 of the General Schedule under section 5332 of such title, and

(2) procure temporary and intermittent services to the same extent as is authorized by section 3109 of title 5, United States Code, but at daily rates for individuals not in excess of the maximum daily rate for GS-18 of the General Schedule under Section 5332 of such title.

(d) The Commission is authorized to enter into contracts with Federal or State agencies, private firms, institutions, and individuals for the conduct of research or surveys, the preparation of reports, and other activities necessary to the discharge of its duties.

COMPENSATION OF MEMBERS

Sec. 6. Members of the Commission (other than members who are officers or employees of the Federal Government) shall receive compensation for each day they are engaged in the performance of their duties as members of the Commission at the rate prescribed for positions at level II of the executive pay schedule in Section 5313 of Title 5, United States Code. Members of the Commission who are officers or employees of the Federal Government shall receive no additional pay on account of their services on the Commission. All members of the Commission shall be entitled to reimbursement for travel, subsistence, and other necessary expenses incurred by them in the performance of their duties as members of the Commission.

APPROPRIATIONS AUTHORIZED

Sec. 7. For the purpose of carrying out this joint resolution, there are authorized to be appropriated such sums as may be necessary, but not to exceed $1,000,000 for each of the two years during which the Commission shall serve.

TERMINATION

Sec. 8. On the ninetieth day after the date of submission of its final report to the President and the Congress, the Commission shall cease to exist.

Summary

The resolution establishes a National Advisory Commission on Health Science and Society, to consist of 15 members appointed by the President. The members would be drawn from the general public and from a variety of disciplines relevant to biomedical research and technology and to the implications thereof.

The commission would make a two-year investigation and study of the ethical, social, and legal implications of advances in biomedical research and technology. After submitting to the President and to the Congress one or more interim reports and a final report, not later than two years after its first meeting, the commission would cease to exist.

APPENDIX 7

"Sex Control, Science and Society" *

Amitai Etzioni

Using various techniques developed as a result of fertility research, scientists are experimenting with the possibility of sex control, the ability to determine whether a newborn infant will be a male or a female. So far, they have reported considerable success in their experiments with frogs and rabbits, whereas the success of experiments with human sperm appears to be quite limited, and the few optimistic reports seem to be unconfirmed. Before this new scientific potentiality becomes a reality, several important questions must be considered. What would be the societal consequences of sex control? If they are, on balance, undesirable, can sex control be prevented without curbing the freedoms essential for scientific work? The scientific ethics already impose some restraints on research to safeguard the welfare and privacy of the researched population. Sex control, however, might affect the whole society. Are there any circumstances under which the societal well-being justifies some limitation on the freedom of research? These questions apply, of course, to many other areas of scientific inquiry, such as work on the biological code and the experimental use of behavior and thought-modifying drugs. Sex control provides a useful opportunity for discussion of these issues because it presents a relatively "low-key" problem. Success seems fairly remote, and, as we

*This paper was first published in *Science*, vol. 161, pages 1107–1112, September 13, 1968. Copyright 1968 by the American Association for the Advancement of Science. I presented an earlier version of this paper to the International Symposium on Science and Politics at Lund, Sweden, June 1968.

shall see, the deleterious effects of widespread sex control would probably not be very great. Before dealing with the possible societal effects of sex control, and the ways they may be curbed, I describe briefly the work that has already been done in this area.

The State of the Art

Differential centrifugation provided one major approach to sex control. It was supposed that since X and Y chromosomes differ in size (Y is considerably smaller), the sperm carrying the two different types would also be of two different weights; the Y-carrying sperm would be smaller and lighter, and the X-carrying sperm would be larger and heavier. Thus, the two kinds could be separated by centrifugation and then be used in artificial insemination. Early experiments, however, did not bear out this theory. And, Witschi pointed out that, in all likelihood, the force to be used in centrifugation would have to be of such magnitude that the sperm may well be damaged (1).

In the 1950s a Swedish investigator, Lindahl (2), published accounts of his results with the use of counterstreaming techniques of centrifugation. He found that by using the more readily sedimenting portion of bull spermatozoa that had undergone centrifugation, fertility was decreased but the number of male calves among the offspring was relatively high. His conclusion was that the female-determining spermatozoa are more sensitive than the male and are damaged due to mechanical stress in the centrifuging process.

Electrophoresis of spermatozoa is reported to have been successfully carried out by a Soviet biochemist, V. N. Schröder, in 1932 (3). She placed the cells in a solution in which the pH could be controlled. As the pH of the solution changed, the sperm moved with different speeds and separated into three groups: some concentrated next to the anode, some next to the cathode, and some were bunched in the middle. In tests conducted by Schröder and N. K. Kolstov (3), sperm which collected next to the anode produced six offspring, all females; those next to the cathode—four males and one female; and those which bunched in the center—two males and two females. Experiments with rabbits over the subsequent 10 years were reported as successful in controlling the sex of the offspring in 80 percent of the cases. Similar success with other mammals is reported.

At the Animal Reproduction Laboratory of Michigan State University, Gordon replicated these findings, although with a lower rate of success *(4)*. Of 167 births studied, in 31 litters, he predicted correctly the sex of 113 offspring, for an average of 67.7 percent. Success was higher for females (62 out of 87, or 71.3 percent) than for males (51 out of 80, or 63.7 percent).

From 1932 to 1942, emphasis in sex control was on the acid-alkali method. In Germany, Unterberger reported in 1932 that in treating women with highly acidic vaginal secretions for sterility by use of alkaline douches, he had observed a high correlation between alkalinity and male offspring. Specifically, over a 10-year period, 53 out of 54 treated females are reported to have had babies, and all of the babies were male. In the one exception, the woman did not follow the doctor's prescription, Unterberger reported *(5)*. In 1942, after repeated tests and experiments had not borne out the earlier results, interest in the acid-alkali method faded *(6)*.

It is difficult to determine the length of time it will take to establish routine control of the sex of animals (of great interest, for instance, to cattle breeders); it is even more difficult to make such an estimate with regard to the sex control of human beings. In interviewing scientists who work on this matter, we heard conflicting reports about how close such a breakthrough was. It appeared that both optimistic and pessimistic estimates were vague—"between 7 to 15 years"—and were not based on any hard evidence but were the researchers' way of saying, "don't know" and "probably not very soon." No specific roadblocks which seemed unusually difficult were cited, nor did they indicate that we have to await other developments before current obstacles can be removed. Fertility is a study area in which large funds are invested these days, and we know there is a correlation between increased investment and findings *(7)*. Although most of the money is allocated to birth control rather than sex-control studies, information needed for sex-control research has been in the past a by-product of the originally sponsored work. Schröder's findings, for example, were an accidental result of a fertility study she was conducting *(4, p.90)*.

Nothing we heard from scientists working in this area would lead one to conclude that there is any specific reason we could not have sex control 5 years from now or sooner.

In addition to our uncertainty about when sex control might be possible, the question of how it would be effected is significant and also one on which there are differences of opinion. The mechanism for practicing sex control is important because certain techniques have greater psychic costs than others. We can see today, for example, that some methods of contraception are preferred by some classes of people because they involve less psychic "discomfort" for them; for example, the intrauterine device is preferred over sterilization by most women. In the same way, although electrophoresis now seems to offer a promising approach to sex control, its use would entail artificial insemination. And whereas the objections to artificial insemination are probably decreasing, the resistance to it is still considerable (8). (Possibly, the opposition to artificial insemination would not be as great in a sex-control situation because the husband's own sperm could be used.) If drugs taken orally or douches could be relied upon, sex control would probably be much less expensive (artificial insemination requires a doctor's help), much less objectionable emotionally, and significantly more widely used.

In any event both professional forecasters of the future and leading scientists see sex control as a mass practice in the foreseeable future. Kahn and Wiener, in their discussion of the year 2000, suggest that one of the "one hundred technical innovations likely in the next thirty-three years" is the "capability to choose the sex of unborn children" (9). Muller takes a similar position about gene control in general (10).

Societal Use of Sex Control

If a simple and safe method of sex control were available, there would probably be no difficulty in finding the investors to promote it because there is a mass-market potential. The demand for the new freedom to choose seems well established. Couples have preferences on whether they want boys or girls. In many cultures boys provide an economic advantage (as workhorses) or as a form of old-age insurance (where the state has not established it). Girls in many cultures are a liability; a dowry which may be a sizeable economic burden must be provided to marry them off. (A working-class American who has to provide for the weddings of three or four daughters may appreciate the problem.) In other cultures, girls are profitably sold. In our own culture, prestige differences are attached to the sex of one's children, which seem to vary among ethnic groups and classes *(11,* pp. 6–7).

Our expectations as to what use sex control might be put in our society are not a matter of idle speculation. Findings on sex preferences are based on both direct "soft" and indirect "hard" evidence. For soft evidence, we have data on preferences parents expressed in terms of the number of boys and girls to be conceived in a hypothetical situation in which parents would have a choice in the matter. Winston studied 55 upperclassmen, recording anonymously their desire for marriage and children. Fifty-two expected to be married some day; all but one of these desired children; expectations of two or three children were common. In total, 86 boys were desired as compared to 52 girls, which amounts to a 65 percent greater demand for males than for females *(12).*

A second study of attitudes, this one conducted on an Indianapolis sample in 1941, found similar preferences for boys. Here, while about half of the parents had no preferences (52.8 percent of the wives and 42.3 percent of the husbands), and whereas the wives with a preference tended to favor having about as many boys as girls (21.8 percent to 25.4 percent), many more husbands wished for boys (47.7 percent as compared to 9.9 percent) *(13).*

Such expressions of preference are not necessarily good indicators of actual behavior. Hence of particular interest is "hard" evidence of what parents actually did—in the limited area of choice they already have: the sex composition of the family at the point they decided to stop having children. Many other and more powerful factors affect a couple's decision to curb further births, and the sex composition of their children is one of them. That is, if a couple has three girls and it strongly desires a boy, this is one reason it will try "once more." By comparing the number of families which had only or mainly girls and "tried once more" to those which had only or mainly boys, we gain some data as to which is considered a less desirable condition. A somewhat different line was followed in an early study. Winston studied 5466 completed families and found that there were 8329 males born alive as compared to 7434 females, which gives a sex ratio at birth of 112.0. The sex ratio of the last child, which is of course much more indicative, was 117.4 (2952 males to 2514 females). That is, significantly more families stopped having children after they had a boy than after they had a girl.

The actual preference for boys, once sex control is available, is likely to be larger than these studies suggest for the following reasons. Attitudes, especially where there is no actual choice, reflect what people believe they ought to believe in, which, in our culture, is equality of the sexes. To prefer to produce boys is lower class and discriminatory. Many middle-class parents might entertain such preferences but be either unaware of them or unwilling to express them to an interviewer, especially since at present there is no possibility of determining whether a child will be a boy or a girl.

Also, in the situations studied so far, attempts to change the sex composition of a family involved having more children than the couple wanted, and the chances of achieving the desired composition were 50 percent or lower. Thus, for instance, if parents wanted, let us say, three children including at least one boy, and they had tried three times and were blessed with girls, they would now desire a boy strongly enough to overcome whatever resistance they had to have additional children before they would try again. This is much less practical than taking a medication which is, let us say, 99.8 percent effective and having the number of children you actually want and are able to support. That is, sex control by a medication is to be expected to be significantly more widely practiced than conceiving more children and gambling on what their sex will be.

Finally, and most importantly, such decisions are not made in the abstract, but affected by the social milieu. For instance, in small *kibbutzim* many more children used to be born in October and November each year than any other months because the community used to consider it undesirable for the children to enter classes in the middle of the school year, which in Israel begins after the high holidays, in October. Similarly, sex control—even if it were taboo or unpopular at first—could become quite widely practiced once it became fashionable.

In the following discussion we bend over backward by assuming that actual behavior would reveal a smaller preference than the existing data and preceding analysis would lead one to expect. We shall assume only a 7 percent difference between the number of boys and girls to be born alive due to sex control, coming on top of the 51.25 to 48.75 existing biological pattern, thus making for 54.75 boys to 45.25 girls, or a surplus of 9.5 boys out of every hundred. This would amount to a surplus of 357,234 in the United States, if sex control were practiced in a 1965-like population *(14)*.

The extent to which such a sex imbalance will cause social dislocations is in part a matter of the degree to which the effect will be cumulative. It is one thing to have an unbalanced baby crop one year, and quite another to produce such a crop several years in a row. Accumulation would reduce the extent to which girl shortages can be overcome by one age group raiding older and younger ones.

Some demographers seem to believe in an invisible hand (as it once was popular to expect in economics), and suggest that overproduction of boys will increase the value of girls and hence increase their production, until a balance is attained under controlled conditions which will be similar to the natural one. We need not repeat here the reasons such invisible arrangements frequently do not work; the fact is they simply cannot be relied upon, as recurrent economic crises in pre-Keynesian days or overpopulation show.

Second, one ought to note the deep-seated roots of the boy-favoring factors. Although there is no complete agreement on what these factors are, and there is little research, we do know that they are difficult and slow to change. For instance, Winston argued that mothers prefer boys as a substitute for their own fathers, out of search for security or Freudian considerations. Fathers prefer boys because boys can more readily achieve success in our society (and in most others). Neither of these factors is likely to change rapidly if the percentage of boys born increases a few percentage points. We do not need to turn to alarmist conclusions, but we ought to consider what the societal effects of sex control might be under conditions of relatively small imbalance which, as we see it, will cause a significant (although not necessarily very high) male surplus, and a surplus which will be cumulative.

Societal Consequences

In exploring what the societal consequences may be, we again need not rely on the speculation of what such a society would be like; we have much experience and some data on societies whose sex ratio was thrown off balance by war or immigration. For example, in 1960 New York City had 343,470 more females than males, a surplus of 68,366 in the 20- to 34-age category alone *(15)*.

We note, first, that most forms of social behavior are sex correlated, and hence that changes in sex composition are very likely to affect most aspects of social life. For instance, women read more books, see more plays, and in general consume more culture than men in the contemporary United States. Also, women attend church more often and are typically charged with the moral education of children. Males, by contrast, account for a much higher proportion of crime than females. A significant and cumulative male surplus will thus produce a society with some of the rougher features of a frontier town. And, it should be noted, the diminution of the number of agents of moral education and the increase in the number of criminals would accentuate already existing tendencies which point in these directions, thus magnifying social problems which are already overburdening our society.

Interracial and interclass tensions are likely to be intensified because some groups, lower classes and minorities specifically *(16)*, seem to be more male-oriented than the rest of the society. Hence while the sex imbalance in a society-wide average may be only a few percentage points, that of some groups is likely to be much higher. This may produce an especially high boy surplus in lower status groups. These extra boys would seek girls in higher status groups (or in some other religious group than their own) *(11)*—in which they also will be scarce.

On the lighter side, men vote systematically and significantly more Democratic than women; as the Republican party has been losing consistently in the number of supporters over the last generation anyhow, another 5-point loss could undermine the two-party system to a point where Democratic control would be uninterrupted. (It is already the norm, with Republicans having occupied the White House for 8 years over the last 36.) Other forms of imbalance which cannot be predicted are to be expected. "All social life is affected by the proportions of the sexes. Wherever there exists a considerable predominance of one sex over the other, in point of numbers, there is less prospect of a well-ordered social life.... Unbalanced numbers inexorably produce unbalanced behavior *(17)*."

Society would be very unlikely to collapse even if the sex ratio were to be much more seriously imbalanced than we expect. Societies are surprisingly flexible and adaptive entities. When asked what would be expected to happen if sex control were available on a mass basis, Davis, the well-known demographer, stated that some delay in the age of marriage of the male, some rise in prostitution and in homosexuality, and some increase in the number of males who will never marry are likely to result. Thus, all of the "costs" that would be generated by sex control will probably not be charged against one societal sector, that is, would not entail only, let us say, a sharp rise in prostitution, but would be distributed among several sectors and would therefore be more readily absorbed. An informal examination of the situation in the USSR and Germany after World War II (sex ratio was 77.7 in the latter) as well as Israel in early immigration periods, support Davis's nonalarmist position. We must ask, though, are the costs justified? The dangers are not apocalyptical; but are they worth the gains to be made?

A Balance of Values

We deliberately chose a low-key example of the effects of science on society. One can provide much more dramatic ones; for example, the invention of new "psychedelic" drugs whose damage to genes will become known only much later (LSD was reported to have such effects), drugs which cripple the fetus (which has already occurred with the marketing of thalidomide), and the attempts to control birth with devices which may produce cancer (early versions of the intrauterine device were held to have such an effect). But let us stay with a finding which generates only relatively small amounts of human misery, relatively well distributed among various sectors, so as not to severely undermine society but only add, maybe only marginally, to the considerable social problems we already face. Let us assume that we only add to the unhappiness of seven out of every 100 born (what we consider minimum imbalance to be generated), who will not find mates and will have to avail themselves of prostitution, homosexuality, or be condemned to enforced bachelorhood. (If you know someone who is desperate to be married but cannot find a mate, this discussion will be less abstract for you; now multiply this by 357,234 per annum.) Actually, to be fair, one must subtract from the unhappiness that sex control almost surely will produce, the joy it will bring to parents who will be able to order the sex of their children; but as of now, this is for most, not an intensely felt need, and it seems a much smaller joy compared to the sorrows of the unmatable mates.

We already recognize some rights of human guinea pigs. Their safety and privacy are not to be violated even if this means delaying the progress of science. The "rest" of the society, those who are not the subjects of research, and who are nowadays as much affected as those in the laboratory, have been accorded fewer rights. Theoretically, new knowledge, the basis of new devices and drugs, is not supposed to leave the inner circles of science before its safety has been tested on animals or volunteers, and in some instances approved by a government agency, mainly the Federal Drug Administration. But as the case of lysergic acid diethylamide (LSD) shows, the trip from the reporting of a finding in a scientific journal to the bloodstream of thousands of citizens may be an extremely short one. The transition did take quite a number of years, from the days in 1943 when Hoffman, one of the two men who synthesized LSD-25 at Sandoz Research Laboratories, first felt its hallucinogenic effect, until the early 1960s, when it "spilled" into illicit campus use. (The trip from legitimate research, its use at Harvard, to illicit unsupervised use was much shorter.) The point is that no additional technologies had to be developed; the distance from the chemical formula to illicit composition required in effect no additional steps.

More generally, Western civilization, ever since the invention of the steam engine, has proceeded on the assumption that society must adjust to new technologies. This is a central meaning of what we refer to when we speak about an industrial revolution; we think about a society being transformed and not just a new technology being introduced into a society which continues to sustain its prior values and institutions. Although the results are not an unmixed blessing (for instance, pollution and traffic casualties), on balance the benefits in terms of gains in standards of living and life expectancy much outweigh the costs. (Whether the same gains could be made with fewer costs if society would more effectively guide its transformation and technology inputs, is a question less often discussed [18].) Nevertheless we must ask, especially with the advent of nuclear arms, if we can expect such a favorable balance in the future. We are aware that single innovations may literally blow up societies or civilization; we must also realize that the rate of social changes required by the accelerating stream of technological innovations, each less dramatic by itself, may supersede the rate at which society can absorb. Could we not regulate to some extent the pace and impact of the technological inputs and select among them without, by every such act, killing the goose that lays the golden eggs?

Scientists often retort with two arguments. Science is in the business of searching for truths, not that of manufacturing technologies. The applications of scientific findings are not determined by the scientists, but by society, politicians, corporations, and the citizens. Two scientists discovered the formula which led to the composition of LSD, but chemists do not determine whether it is used to accelerate psychotherapy or to create psychoses, or, indeed, whether it is used at all, or whether, like thousands of other studies and formulas, it is ignored. Scientists split the atom, but they did not decide whether particles would be used to produce energy to water deserts or superbombs.

Second, the course of science is unpredictable, and any new lead, if followed, may produce unexpected bounties; to curb some lines of inquiry—because they may have dangerous outcomes—may well force us to forego some major payoffs; for example, if one were to forbid the study of sex control one might retard the study of birth control. Moreover, leads which seem "safe" may have dangerous outcomes. Hence, ultimately, only if science were stopped altogether, might findings which are potentially dangerous be avoided.

These arguments are often presented as if they themselves were empirically verified or logically true statements. Actually they are a formula which enables the scientific community to protect itself from external intervention and control. An empirical study of the matter may well show that science does thrive in societies where scientists are given less freedom than the preceding model implies science must have for example, in the Soviet Union. Even in the West in science some limitations on work are recognized and the freedom to study is not always seen as the ultimate value. Whereas some scientists are irritated when the health or privacy of their subject curbs the progress of their work, most scientists seem to recognize the priority of these other considerations. (Normative considerations also much affect the areas studied; compare, for instance, the high concern with a cancer cure to the almost complete unwillingness of sociologists, since 1954, to retest the finding that separate but equal education is not feasible.)

One may suggest that the society at large deserves the same protection as human subjects do from research. That is, the scientific community cannot be excused from the responsibility of asking what effects its endeavors have on the community. On the contrary, only an extension of the existing codes and mechanisms of self-control will ultimately protect science from a societal backlash and the heavy hands of external regulation. The intensification of the debate over the scientists' responsibilities with regard to the impacts of their findings is by itself one way of exercising it, because it alerts more scientists to the fact that the areas they choose to study, the ways they communicate their findings (to each other and to the community), the alliances they form or avoid with corporate and governmental interests—all these affect the use to which their work is put. It is simply not true that a scientist working on cancer research and one working on biological warfare are equally likely to come up with a new weapon and a new vaccine. Leads are not that random, and applications are not that readily transferable from one area of application to another.

Additional research on the societal impact of various kinds of research may help to clarify the issues. Such research even has some regulatory impact. For instance, frequently when a drug is shown to have been released prematurely, standards governing release of experimental drugs to mass production are tightened (19), which in effect means fewer, more carefully supervised technological inputs into society; at least society does not have to cope with dubious findings. Additional progress may be achieved by studying empirically the effects that various mechanisms of self-regulation actually have on the work of scientists. For example, urging the scientific community to limit its study of some topics and focus on others may not retard science; for instance, sociology is unlikely to suffer from being now much more reluctant to concern itself with how the U.S. Army may stabilize or undermine foreign governments than it was before the blowup of Project Camelot (20).

In this context, it may be noted that the systematic attempt to bridge the "two cultures" and to popularize science has undesirable side effects which aggravate the problem at hand. Mathematical formulas, Greek or Latin terminology, and jargon were major filters which allowed scientists in the past to discuss findings with each other without the nonprofessionals listening in. Now, often even preliminary findings are reported in the mass media and lead to policy adaptations, mass use, even legislation (21), long before scientists have had a chance to double-check the findings themselves and their implications. True, even in the days when science was much more esoteric, one could find someone who could translate its findings into lay language and abuse it; but the process is much accelerated by well-meaning men (and foundations) who feel that although science ought to be isolated from society, society should keep up with science as much as possible. Perhaps the public relations efforts on behalf of science ought to be reviewed and regulated so that science may remain free.

A system of regulation which builds on the difference between science and technology, with some kind of limitations on the technocrats serving to protect societies, coupled with little curbing of scientists themselves, may turn out to be much more crucial. The societal application of most new scientific findings and principles advances through a sequence of steps, sometimes referred to as the R & D process. An abstract finding or insight frequently must be translated into a technique, procedure, or hardware, which in turn must be developed, tested, and mass-produced, before it affects society. While in some instances, like that of LSD, the process is extremely short in that it requires few if any steps in terms of further development of the idea, tools, and procedures, in most instances the process is long and expensive. It took, for instance, about $2 billion and several thousand applied scientists and technicians to make the first atomic weapons after the basic principles of atomic fission were discovered. Moreover, technologies often have a life of their own; for example, the intrauterine device did not spring out of any application of a new finding in fertility research but grew out of the evolution of earlier technologies.

The significance of the distinction between the basic research ("real" science) and later stages of research is that, first, the damage caused (if any) seems usually to be caused by the technologies and not by the science applied in their development. Hence if there were ways to curb damaging technologies, scientific research could maintain its almost absolute, follow-any-lead autonomy and society would be protected.

Second, and most important, the norms to which applied researchers and technicians subscribe and the supervisory practices, which already prevail, are very different than those which guide basic research. Applied research and technological work are already intensively guided by societal, even political, preferences. Thus, while about $2 billion a year of R & D money are spent on basic research more or less in ways the scientists see fit, the other $13 billion or so are spent on projects specifically ordered, often in great detail, by government authorities, for example, the development of a later version of a missile or a "spiced-up" tear gas. Studies of R & D corporations—in which much of this work is carried out, using thousands of professionals organized in supervised teams which are given specific assignments—pointed out that wide freedom of research simply does not exist here. A team assigned to cover a nose cone with many different alloys and to test which is the most heat-resistant is currently unlikely to stumble upon, let us say, a new heart pump, and if it were to come upon almost any other lead, the boss would refuse to allow the team to pursue the lead, using the corporation's time and funds specifically contracted for other purposes.

Not only are applied research and technological developments guided by economic and political considerations but also there is no evidence that they suffer from such guidance. Of course, one can overdirect any human activity, even the carrying of logs, and thus undermine morale, satisfaction of the workers, and their productivity; but such tight direction is usually not exercised in R & D work nor is it required for our purposes. So far guidance has been largely to direct efforts toward specific goals, and it has been largely corporate, in the sense that the goals have been chiefly set by the industry (for example, building flatter TV sets) or mission-oriented government agencies (for instance, hit the moon before the Russians). Some "preventive" control, like the suppression of run-proof nylon stockings, is believed to have taken place and to have been quite effective.

I am not suggesting that the direction given to technology by society has been a wise one. Frankly, I would like to see much less concern with military hardware and outer space and much more investment in domestic matters; less in developing new consumer gadgets and more in advancing the technologies of the public sector (education, welfare, and health); less concern with nature and more with society. The point though is that, for good or bad, technology is largely already socially guided, and hence the argument that its undesirable effects cannot be curbed because it cannot take guidance and survive is a false one.

What may have to be considered now is a more preventive and more national effective guidance, one that would discourage the development of those technologies which, studies would suggest, are likely to cause significantly more damage than payoffs. Special bodies, preferably to be set up and controlled by the scientific community itself, could be charged with such regulation, although their decrees might have to be as enforceable as those of the Federal Drug Administration. (The Federal Drug Administration, which itself is overworked and understaffed, deals mainly with medical and not societal effects of new technologies.) Such bodies could rule, for instance, that whereas fertility research ought to go on uncurbed, sex-control procedures for human beings are not to be developed.

One cannot be sure that such bodies would come up with the right decisions. But they would have several features which make it likely that they would come up with better decisions than the present system for the following reasons: (i) they would be responsible for protecting society, a responsibility which so far is not institutionalized; (ii) if they act irresponsibly, the staff might be replaced, let us say by a vote of the appropriate scientific associations; and (iii) they would draw on data as to the societal effects of new (or anticipated) technologies, in part to be generated at their initiative, while at present—to the extent such supervisory decisions are made at all—they are frequently based on folk knowledge.

Most of us recoil at any such notion of regulating science, if only at the implementation (or technological) end of it, which actually is not science at all. We are inclined to see in such control an opening wedge which may lead to deeper and deeper penetration of society into the scientific activity. Actually, one may hold the opposite view—that unless societal costs are diminished by some acts of self-regulation at the stage in the R & D process where it hurts least, the society may "backlash" and with a much heavier hand slap on much more encompassing and throttling controls.

The efficacy of increased education of scientists to their responsibilities, of strengthening the barriers between intrascientific communications and the community at large, and of self-imposed, late-phase controls may not suffice. Full solution requires considerable international cooperation, at least among the top technology-producing countries. The various lines of approach to protecting society discussed here may be unacceptable to the reader. The problem though must be faced, and it requires greater attention as we are affected by an accelerating technological output with ever-increasing societal ramifications, which jointly may overload society's capacity to adapt and individually cause more unhappiness than any group of men has a right to inflict on others, however noble their intentions.

References and Notes

1. E. Witschi, personal communication.
2. P. E. Lindahl, *Nature* 181, 784 (1958).
3. V. N. Schröder and N. K. Koltsov, *ibid.* 131, 329 (1933).
4. M. J. Gordon, *Sci. Amer.* 199, 87–94 (1958).
5. F. Unterberger, *Deutsche Med. Wochenschr.* 56, 304 (1931).
6. R. C. Cook, *J. Hered.* 31, 270 (1940).
7. J. Schmookler, *Invention and Economic Growth* (Harvard Univ. Press, Cambridge, Mass., 1966).
8. Many people prefer adoption to artificial insemination. See G. M. Vernon and J. A. Boadway, *Marriage Family Liv.* 21, 43 (1959).
9. H. Kahn and A. J. Wiener, *The Year 2000: A Framework for Speculation on the Next Thirty-Three Years* (Macmillan, New York, 1967), p. 53.
10. H. J. Muller, *Science* 134, 643 (1961).
11. C. F. Westoff, "The social-psychological structure of fertility," in *International Population Conference* (International Union for Scientific Study of Population, Vienna, 1959).
12. S. Winston, *Amer. J. Sociol.* 38, 226 (1932). For a cricial comment which does not affect the point made above, see H. Weiler, *ibid.* 65, 298 (1959).
13. J. E. Clare and C. V. Kiser, *Millbank Mem. Fund Quart.* 29, 441 (1951). See also D. S. Freedman, R. Freedman, P. K. Whelpton, *Amer. J. Sociol.* 66, 141 (1960).
14. Based on the figure for 1965 registered births (adjusted for those unreported) of

3,760,358 from *Vital Statistics of the United States 1965* (U.S. Government Printing Office, Washington, D.C., 1965), vol. 1, pp. 1–4, section 1, table 1–2. If there is a "surplus" of 9.5 boys out of every hundred, there would have been 3,760,358/100 x 9.5 = 357,234 surplus in 1965.

15. Calculated from C. Winkler, Ed., *Statistical Guide 1965 for New York City* (Department of Commerce and Industrial Development, New York, 1965), p. 17.

16. Winston suggests the opposite but he refers to sex control produced through birth control which is more widely practiced in higher classes, especially in the period in which his study was conducted, more than a generation ago.

17. Quoted in J. H. Greenberg, *Numerical Sex Disproportion: A Study in Demographic Determinism* (Univ. of Colorado Press, Boulder, 1950), p. 1. The sources indicated are A. F. Weber, *The Growth of Cities in the Nineteenth Century*, Studies in History, Economics, and Public Law, vol. 11, p. 85, and H. von Hentig, *Crime: Causes and Conditions* (McGraw-Hill, New York, 1947), p. 121.

18. For one of the best discussions, see E. E. Morison, *Men, Machines, and Modern Times* (M.I.T. Press, Cambridge, Mass., 1966). See also A. Etzioni, *The Active Society: A Theory of Societal and Political Processes* (Free Press, New York, 1968), chaps. 1 and 21.

19. See reports in The *New York Times:* "Tranquilizer is put under U.S. curbs; side effects noted," 6 December 1967; "F.D.A. is studying reported reactions to arthritis drug," 19 March 1967; "F.D.A. adds 2 drugs to birth defect list," 3 January 1967. On 24 May 1966, Dr. S. F. Yolles, director of the National Institute of Mental Health, predicted in testimony before a Senate subcommittee: "The next 5 to 10 years . . . will see a hundredfold increase in the number and types of drugs capable of affecting the mind."

20. I. L. Horowitz, *The Rise and Fall of Project Camelot* (M.I.T. Press, Cambridge, Mass., 1967).

21. For a detailed report, see testimony by J. D. Cooper, on 28 February 1967, before the subcommittee on government research of the committee on government operations, United States Senate, 90th Congress (First session on Biomedical Development, Evaluation of Existing Federal Institutions), pp. 46–61.

APPENDIX 8

Ethical and Social Issues in Screening for Genetic Disease*

A report from the Research Group on Ethical, Social and Legal Issues in Genetic Counseling and Genetic Engineering of the Institute of Society, Ethics and the Life Sciences; Marc Lappé, Ph.D., Program Director; and James M. Gustafson, Ph.D., and Richard Roblin, Ph.D., Co-chairmen

Abstract. The potential advent of widespread genetic screening raises new and often unanticipated ethical, psychologic and sociomedical problems for which physicians and the public may be unprepared. To focus attention on the problems of stigmatization, confidentiality, and breaches of individual rights to privacy and freedom of choice in childbearing, we have proposed a set of principles for guiding the operation of genetic screening programs. The main principles emphasized include the need for well planned program objectives, involvement of the communities immediately affected by screening, provision of equal access, adequate testing procedures, absence of compulsion, a well defined procedure for

*The statement represents a summary of the major preliminary and provisional findings of the group. The following are the signatories of the report: James M. Gustafson, Ph.D., School of Religious Studies, Yale University, and Richard Roblin, Ph.D., Infectious Disease Unit, Massachusetts General Hospital (Co-chairmen); Alex Capron, LL.B., Yale Law School; Arthur J. Dyck, Ph.D., Center for Population Studies, Harvard University; Lee Ehrman, Ph.D., Division of Natural Sciences, State University of New York at Purchase; Richard Erbe, M.D., Genetics Unit, Massachusetts General Hospital; John C. Fletcher, Ph.D., Interfaith Me-

obtaining informed consent, safeguards for protecting subjects, open access of communities and individuals to program policies, provision of counseling services, an understanding of the relation of screening to realizable or potential therapies, and well formulated procedures for protecting the rights of individual and family privacy.

In recent months a number of large-scale genetic screening programs for sickle-cell trait and sickle-cell anemia, and at least one for the carrier state in Tay-Sachs disease, have been initiated. Further proliferation of genetic screening programs for these and other genetic diseases seems likely, and in some cases participation in these programs may be made compulsory by statute.* Since screening programs acquire genetic information from large numbers of normal and asymptomatic (e.g., carrier state) individuals and families, often after only brief medical contact, their operation generally falls outside the usual patient-initiated doctor-patient relation. As a result, traditional applications of ethical guidelines for confidentiality and individual physician responsibility are uncertain in mass screening programs. Thus, we believe it important that attempts be made now to clarify some ethical,

---

tropolitan Theological Education, Inc.; Harold P. Green, J.D., National Law Center, George Washington University; Kurt Hirschhorn, M.D., Mount Sinai School of Medicine, New York; Hans Jonas, Ph.D., D.H.L. (honoris causa), New School for Social Research; Michael Kaback, M.D., Department of Pediatrics, Johns Hopkins Medical School; Karen Lebacqz, Harvard University; Ernst Mayr, Ph.D., Museum of Comparative Zoology, Harvard University; William J. Mellman, M.D., Department of Pediatrics and Medical Genetics, University of Pennsylvania Medical School; Arno G. Motulsky, M.D., departments of Medicine and Genetics, University of Washington School of Medicine; Robert F. Murray, Jr., M.D., Department of Pediatrics, Howard University College of Medicine; John Rainer, M.D., New York State Psychiatric Institute; Paul Ramsey, Ph.D., Department of Religion, Princeton University; James R. Sorenson, Ph.D., Department of Sociology, Princeton University; Sumner Twiss, Ph.D., Department of Religious Studies, Brown University; the Institute Staff consists of Marc Lappé, Ph.D., Program Director; Daniel Callahan, Ph.D., Robert M. Veath, Ph.D., Associate for Medical Ethics; and Sharmon Sollitto, Research Associate for Genetics.

New England Journal of Medicine 286 (May 25, 1972), pp. 1129–1132. Reprinted by permission of the authors and the *New England Journal of Medicine*.

*Massachusetts approved an act (Chapter 491 of Acts and Resolves, 1971) on July 1, 1971, "requiring the testing of blood for sickle trait or anemia as a prerequisite to school attendance."

social and legal questions concerning the establishment and operation of such programs. Although we recognize that there are deep divisions regarding the morality of abortion and that certain views would question prenatal diagnosis so far as it involves abortion, we shall not discuss these issues here. In what follows, we have considered the goals that genetic screening programs may serve and have described some principles that we believe are essential to their proper operation.

Goals Served by Screening

It is crucial that screening programs be structured on the basis of one or more clearly identified goals and that such goals be formulated well before screening actually begins. We believe it will prove costly in scientific and human terms to omit or defer a careful evaluation of program objectives. Although there are three distinguishable categories of goals that screening programs may serve, we believe the most important goals are those that either contribute to improving the health of persons who suffer from genetic disorders, or allow carriers for a given variant gene to make informed choices regarding reproduction, or move toward alleviating the anxieties of families and communities faced with the prospect of serious genetic disease. The following are representative statements of goals that have been used to justify screening programs.

THE PROVISION OF BENEFITS TO INDIVIDUALS AND FAMILIES

Such benefits may arise from enabling couples found by screening to be at risk for transmitting a genetic disease to take genetic information into account in making responsible decisions about having or not having children. This usually is done by providing genetic counseling services and informing couples about the nature of existing alternatives and potential therapies (e.g., sickle-cell screening). Another advantage consists in detecting asymptomatic persons at birth when amelioration of the sequelae of a genetic disease is already possible—e.g., screening for phenylketonuria (PKU). Still another is providing means for couples, found at risk by screening, to have children free from a severe and untreatable genetic disease (e.g., Tay-Sachs screening).

ACQUISITION OF KNOWLEDGE ABOUT GENETIC DISEASE

Laboratory research and theoretical studies have had a major role in helping to understand fundamental aspects of human genetic diseases. In addition, however, some large-scale screening programs may be needed to determine frequencies of rare diseases and to establish new correlations between genes or groups of genes and disease. In some such screening programs, no therapy may be immediately available for the pathologic condition, although the information derived from them may lead to therapeutic benefits in the future. Research programs aimed primarily at the acquisition of genetic knowledge *per se* are important. Yet we believe their value is enhanced when they also contribute information that is useful for counseling individuals or for public-health purposes.

REDUCTION OF THE FREQUENCY OF APPARENTLY DELETERIOUS GENES

Although little is known about the possible beneficial (or detrimental) effects of most deleterious recessive genes in the heterozygous state, the reduction of their frequency would be one way to decrease the occurrence of suffering caused by their homozygous manifestations. Nevertheless, as a goal of screening programs, the means required to approach this objective appear to be both practically and morally unacceptable. Virtually everyone carries a small number of deleterious or lethal recessive genes, and to reduce the frequency of a particular recessive gene to near the level maintained by recurrent mutation, most or all persons heterozygous for that gene would have either to refrain from procreation entirely or to monitor all their offspring in utero and abort not only affected homozygote fetuses but also the larger number of heterozygote carriers for the gene.[1-3] However, substantial reduction in the frequency of a recessive disease is possible by prenatal screening and selective abortion, or by counseling persons with the same trait to refrain from marriage or childbearing.[3] Nevertheless, these means of reducing the suffering concomitant to recessive disease raise moral questions of their own.

Principles for the Design and Operation of Screening Programs

ATTAINABLE PURPOSE

Before a program is undertaken, planners should have ascertained through pilot projects and other studies that the program's purposes are attainable. Articulating attainable purposes is necessary if the program is to avoid promising (or seeming to promise) results or benefits that it cannot deliver. It is also desirable to update program design and objectives continually in the light of the program experience and new medical developments. Consideration might also be given to incorporating additional purposes—for example, sickle-cell screening programs might profitably enlarge their scope to include other hemoglobinopathies[4] as well as general screening for anemia.[5]

COMMUNITY PARTICIPATION

From the outset program planners should involve the communities affected by screening in formulating program design and objectives, in administering the actual operation of the program, and in reviewing results. This involvement may include the lay, religious and medical communities as in the Baltimore Tay-Sachs program.[6] Considerable effort should be expended to make program objectives clear to the public, and to encourage participation. Recent articles describing detection programs for Tay-Sachs-disease heterozygotes[6] and for persons with sickle-cell trait or disease[7] have stressed the educational aspect of program design as the crucial component of successful operation. The principal value of community participation is to afford individuals knowledge of the availability and self-determination in the choice of this type of medical service. Educated community involvement is also a means of reducing the potential risk that those identified as genetically variant will be stigmatized or ostracized socially.

EQUAL ACCESS

Information about screening and screening facilities should be open and available to all. To make testing most useful for certain conditions, priority should be given to informing certain well defined populations in which the condition occurs with definitely greater frequency, such as hemoglobin S in blacks and deficient hexosaminidase A (Tay-Sachs disease) among Ashkenazi Jews.

ADEQUATE TESTING PROCEDURES

To avoid the problems that occurred initially in PKU screening,[8] testing procedures should be accurate, should provide maximal information, and should be subject to minimum misinterpretation. For detection of autosomal recessive conditions like sickle-cell anemia, for example, the test used should accurately distinguish between those carrying the trait and those homozygous for the variant gene.[4, 9]

ABSENCE OF COMPULSION

As a general principle, we strongly urge that no screening program have policies that would in any way impose constraints on childbearing by individuals of any specific genetic constitution, or would stigmatize couples who, with full knowledge of the genetic risks, still desire children of their own. It is unjustifiable to promulgate standards for normalcy based on genetic constitution. Consequently, genetic screening programs should be conducted on a voluntary basis. Although vaccination against contagious diseases and premarital blood tests are sometimes made mandatory to protect the public health, there is currently no public-health justification for mandatory screening for the prevention of genetic disease. The conditions being tested for in screening programs are neither "contagious" nor, for the most part, susceptible to treatment at present.[10]

INFORMED CONSENT

Screening should be conducted only with the informed consent of those tested or of the parents or legal representatives of minors. We seriously question the rationale of screening preschool minors or preadolescents for sickle-cell disease or trait since there is a substantial danger of stigmatization and little medical value in detecting the carrier state at this age. However, in the light of recent information that sickle-cell crises can potentially be mitigated,[10] a beneficial alternative would be newborn screening that could identify the SS homozygote in early life, and thereby anticipate the problems and complications associated with sickle-cell disease and provide early counseling to the parents.

In addition to obtaining signed consent documents, it is the program director's obligation to assure that knowledgeable consent is obtained from all those screened, to design and implement informational procedures, and to review the consent procedure for its effectiveness. The guidelines available from the Department of Health, Education, and Welfare[11] provide a useful model for formulating such consent procedures.

PROTECTION OF SUBJECTS

Since genetic screening is generally undertaken with relatively untried testing procedures[9] and is vitally concerned with the acquisition of new knowledge, it ought properly to be considered a form of "human experimentation." Although most screening entails only minimum physical hazard for the participants, there is a risk of possible psychologic or social injury, and screening programs should consequently be conducted according to the guidelines set forth by HEW for the protection of research subjects.[11]

ACCESS TO INFORMATION

A screening program should fully and clearly disclose to the community and all persons being screened its policies for informing those screened of the results of the tests performed on them. As a general rule all unambiguous diagnostic results should be made available to the person, his legal representative, or a physician authorized by him. Where full disclosure is not practiced, the burden of justifying nondisclosure lies with those who would withhold information. If an adequate educational program has been offered on the meaning of diagnostic criteria and subjects participate in the screening voluntarily, it may generally be assumed that they are emotionally prepared to accept the information derived from the testing.

PROVISION OF COUNSELING

Well-trained genetic counselors should be readily available to provide adequate assistance (including repeated counseling sessions if necessary) for persons identified as heterozygotes or more rarely homozygotes by the screening program. As a general rule, counseling should be nondirective, with an emphasis on informing the client and not making decisions for him.[12] The need for defining appropriate qualifications for genetic counselors in the context of screening programs and for providing adequate numbers of trained counselors remains an urgent one. It is the ongoing responsibility of the program directors to evaluate the effectiveness of their program by follow-up surveys of their counseling services. This may include steps (taken with the prior understanding and approval of the subjects screened) to determine how well the information about genetic status has been understood and how it has affected the participants' lives.

UNDERSTANDABLE RELATION TO THERAPY

As part of the educational process that precedes the actual testing program, the nature and cost of available therapies or maintenance programs for affected offspring, combined with an understandable description of their possible benefits and risks, should be given to all persons to be screened. We believe this is one of the items of information that subjects need in deciding whether or not to participate in the program. In addition, acceptance of research therapy should not be a precondition for participation in screening, nor should acceptance of screening be construed as tacit acceptance of such therapy. Both those doing the testing and those doing the counseling ought to keep abreast of existing and imminent developments in diagnosis and therapy[10,13-15] so that the goals of the program and information offered to those being screened will be consistent with the therapeutic options available.

PROTECTION OF RIGHT OF PRIVACY

Well-formulated procedures should be set up in advance of actual screening to protect the rights of privacy of individuals and their families. We note that the majority of states do not have statutes that recognize the confidentiality of public-health information or are even minimally adequate to protect individual privacy.[16] Researchers therefore have a particularly strong obligation to protect screening information. Consequently, we favor policies of informing only the person to be screened or, with his permission, a designated physician or medical facility, of having records kept in code, of prohibiting storage of noncoded information in data banks where telephone computer access is possible and of limiting private and public access only to anonymous data to be used for statistical purposes.

Conclusions

Even if the above guidelines are followed, some risk will remain that the information derived from genetic screening will be misused. Such misuse or misinterpretation must be seen as one of the principal potentially deleterious consequences of screening programs. Several medical researchers have recently cautioned their colleagues of the potential for misinterpretation of the clinical meaning of sickle "trait" and "disease."[5] We are concerned about the dangers of societal misinterpretation of similar conditions and the possibility of widespread and undesirable labeling of individuals on a genetic basis. For instance, the lay public may incorrectly conclude that persons with sickle trait are seriously handicapped in their ability to function effectively in society. Moreover, protecting the confidentiality of test results will not shield all such subjects from a felt sense of stigmatization nor from personal anxieties stemming from their own misinterpretation of their carrier status. Extreme caution should therefore be exercised before steps that lend themselves to stigmatization are taken—for example, stigmatization can arise from recommending restrictions on young children's physical activities under normal conditions because of sickle-cell trait, or from denying life-insurance coverage to adult trait carriers, neither of which are currently medically indicated. In view of such collateral risks of screening, it is essential that each program's periodic review include careful consideration of the social and psychologic ramifications of its operation.

References

1. Crow JF: Population perspective, Ethical Issues in Genetic Counseling and the Use of Genetic Knowledge. Edited by P Condliffe. D Callahan, B Hilton, et al. New York, Plenum Press (in press)

2. Morton NE: Population genetics and disease control. Soc Biol 18: 243–251, 1971

3. Motulsky AG, Frazier GR, Felsenstein J: Public health and long-term genetic implications of intrauterine diagnosis and selective abortion, Intrauterine Diagnosis (Birth Defects Original Article Series Vol 7, No 5). Edited by D Bergsma. New York, The National Foundation, 1971, pp. 22–32

4. Barnes MG, Komarmy L, Novack AH: A comprehensive screening program for hemoglobinopathies. JAMA 219:701–705, 1972

5. Beutler E, Boggs DR, Heller P, et al: Hazards of indiscriminate screening for sickling. N Engl J Med 285:1485–1486, 1971

6. Kaback MM, Zieger RS: The John F. Kennedy Institute Tay Sachs program: practical and ethical issues in an adult genetic screening program, Ethical Issues in Genetic Counseling and the Use of Genetic Knowledge. Edited by P Condliffe, D Callahan, B Hilton, et al. New York. Plenum Press (in press)

7. Nalbandian RM, Nichols BM, Heustis AE, et al: An automated mass screening program for sickle cell disease. JAMA 218:1680–1682, 1971

8. Bessman PS, Swazey JP: PKU: a study of biomedical legislation, Human Aspects of Biomedical Innovation. Edited by E Mendelsohn, JP Swazey, I Taviss. Cambridge, Harvard University Press, 1971, pp 49–76

9. Nalbandian RM, Henry RL, Lusher JM, et al: Sickledex test for hemoglobin S: a critique. JAMA 218:1679–1680, 1971

10. May A, Bellingham AJ, Huehns ER: Effect of cyanate on sickling. Lancet 1:658–661, 1972

11. The Institutional Guide To DHEW Policy on Protection of Human Subjects, Grants Administration Manual, Chapter 1–40. (DHEW Publication No [NIH] 72–102). Washington, DC. Division of Research Grants, Department of Health, Education, and Welfare, 1971

12. Sorenson JR: Social Aspects of Applied Human Genetics (Social Science Frontiers No 3). New York, Russell Sage Foundation, 1971

13. McCurdy PR, Mahmood L: Intravenous urea treatment of the painful crisis of sickle-cell disease: a preliminary report. N Engl J Med 285:992–994, 1971

14. Gillette PN, Manning JM, Cerami A: Increased survival of sickle-cell erythrocytes after treatment in vitro with sodium cyanate. Proc Natl Acad Sci USA 68:2791–2793, 1971

15. Hollenberg MD, Kaback MM, Kazazian HH Jr: Adult hemoglobin synthesis by reticulocytes from the human fetus at midtrimester, Science 174:698–702, 1971

16. Schwitzgebel RB: Confidentiality of research information in public health studies. Harv Leg Comment 6:187–197, 1969

Notes

Preface

1. *See* "Genetic Engineering: Evolution of a Technological Issue." Report to the Subcommittee on Science, Research, and Development of the Committee of Science and Astronautics, U.S. House of Representatives, Ninety-Second Congress (Washington, D.C.: U.S. Government Printing Office, 1972), p. 5. *See also* Theodore Friedmann, "Prenatal Diagnosis of Genetic Disease," *Scientific American* 225 (November 1971), p. 34.
2. Amitai Etzioni, *The Active Society* (New York: Free Press, 1968).

Introduction

1. *Uniform Crime Reports—1970* (Washington, D.C.: U.S. Government Printing Office), p. 129.

Chapters

CHAPTER ONE

1. James S. Coleman, Ernest Q. Campbell, Carol J. Hobson, James McPartland, Alexander M. Mood, Frederick D. Weinfeld, and Robert L. York, *Equality of Educational Opportunity* (Washington, D.C.: U.S. Government Printing Office. 1966).

2. Arthur R. Jensen, "How Much Can We Boost IQ and Scholastic Achievement?" *Harvard Educational Review* 39 (1969), pp. 1–123; Richard J. Herrnstein, "IQ," *The Atlantic* 228 (September 1971), pp. 43–64; Jerrold K. Footlick, "Jensen for the Defense," *Newsweek*, March 19, 1973.

3. Eliot Slater and Valerie Cowie, *The Genetics of Marital Disorders* (New York: Oxford University Press, 1971); "X Marks the Manic," *Newsweek*, August 28, 1972; "Hereditary Drunkenness?" *Newsweek*, April 10, 1972; Earl Ubell, "Genes May Be the Villains," *New York Times*, August 27, 1972; Irving Gottesman and James Shields, *Schizophrenia and Genetics* (New York: Academic Press, 1972).

4. Nathan Glazer, "Paradoxes of Health Care," *The Public Interest* 22 (Winter 1971), pp. 62–77; Irving Kristol, "About Equality," *Commentary* 54 (November 1972), pp. 41–47.

5. Warren G. Bennis and Phillip E. Slater, *The Temporary Society* (New York: Harper & Row, 1968).

6. Anonymous M.D., *Confessions of a Gynecologist* (New York: Doubleday & Co., 1972), p. 224.

7. The following references are for some of the studies that have been done in this area: Kenneth Burke, "The 'XYY Syndrome': Genetics, Behavior and the Law," *Denver Law Journal* 46 (1969), pp. 261–284; Mary A. Telfer, David Baker, Gerald Clark, and Claude Richardson, "Incidence of Gross Chromosomal Errors Among Tall Criminal American Males," *Science* CLIX (March 15, 1968), pp. 1249–1250.

8. Burke, "The XYY Syndrome," pp. 261–284; Telfer et al., "Incidence of Gross Chromosomal Errors," pp. 1249–1250.

9. P.S. Bessman and Judith Swazey, "PKU—A Study of Biomedical Legislation," in E. Mendelson, J. Swazey, and I. Taviss, eds., *Human Aspects of Biomedical Innovation* (Cambridge, Mass.: Harvard University Press, 1971), pp. 49–76. The mass-screening programs are founded on the Guthrie blood tests. These test for high blood phenylalanine levels, which are alone taken as symptoms for the disease. According to Bessman and Swazey, the only clear predictor of PKU is absence of the

protein phenylalanine hydorxylase in the liver, but since a liver biopsy, a complicated and costly technique, is the only means of determining this, the simpler Guthrie test is used. There is evidence, however, that denies an explicit correlation between the excess acid tested for and retardation and severity of retardation. Thus the tests may or may not correctly identify patients. Some cases of PKU pass unidentified, while some normal cases yield "false positive" reactions. The high acid level is not a necessary and sufficient condition for PKU, yet all treatment is based on its presence or absence.

10. R. Guthrie and S. Whitney, reported in Joseph Cooper, "Creative Pluralism, Medical Ombudsman," Research in the Service of Man. Hearings before the Subcommittee on Government Research of the Senate Committee on Government Operations, 90th Congress, 1967, vol. 62, p. 50. For data on the troubles of the tests and related discussion, *see* Cooper, "Creative Pluralism, Medical Ombudsman," pp. 46–66; Bessman and Swazey, "PKU—A Study of Biomedical Legislation," pp. 49–76; "After Ten Years of PKU Testing, a Re-evaluation," *Medical World News,* November 19, 1971, pp. 43–44.

11. Charles Blumenfeld, reported in Cooper, "Creative Pluralism, Medical Ombudsman," p. 51.

12. For example, *see* Lawrence K. Altman, "Artificial Kidney Use Poses Awesome Questions," *New York Times,* October 24, 1971.

13. *See* Jane E. Brody, "How to Save Lives With a 'Registry,'" *New York Times,* September 19, 1971; Gloria Gonzales, "Computer Matches Kidneys in Newark Transplants," *New York Times,* April 16, 1972.

14. "Most Doctors and Teachers Are Not Equipped to Help," *Life Magazine* (October 1972); Alan Charles, "The Case of Ritalin," *New Republic* 165 (October 23, 1971), pp. 17–19; John Else and Alvin Ross, "A Communication," *New Republic* 164 (April 17, 1971), pp. 37–39; Robert Reinhold, "Rx For Child's Learning Malady," *New York Times,* July 3, 1970.

15. Erving Goffman, *Asylums: Essays on the Social Situation of Mental Patients and Other Inmates* (New York: Doubleday & Co., 1961); Ronald Leifer, M.D., *In the Name of Mental Health: The Social Functions of Psychiatry* (New York: Science House, 1969); Thomas Szasz, M.D., *Ideology and Insanity: Essays on the Psychiatric Dehumanization of Man* (New York: Doubleday & Co., 1970).

16. Richard James, "Genetic Blueprint: Horror or Hope?" *Wall Street Journal,* November 11, 1968.

17. James V. McConnell, Richard L. Cutler, and Elton B. McNeil, "Subliminal Stimulation: An Overview," *American Psychologist* 13 (1958), pp. 229–42; Donald P. Spence, "Subliminal Perception and

Perceptual Defense; 2 Sides of a Single Problem," *Behavioral Science* 12 (1967), pp. 183–193; Luther B. Jennings and Stephen G. George, "The Spence-Holland Theory of Subliminal Perception: A Reexamination," *Psychological Record* 20 (1970), pp. 495–504.

18. Ronald M. Costell, Donald T. Lunde, Bert S. Kopell, and William K. Wittner, "Contingent Negative Variation as an Indicator of Sexual Object Preference," *Science* 177 (August 25, 1972), pp. 718–720.

19. Etzioni, *The Active Society.* A book has been recently published that makes *The Active Society* more comprehensible to a less scholarly audience: Warren Breed, *The Self-Guiding Society* (New York: Free Press, 1971). For a reader which includes a number of articles about this theory, *see* Sarajane Heidt and Amitai Etzioni, eds., *Societal Guidance* (New York: Thomas Y. Crowell Co., 1969).

20. For other metaphysical and/or ethical positions on eugenics, *see* Paul Ramsey, "Moral and Religious Implications of Genetic Control," in John D. Roslansky, ed., *Genetics and the Future of Man* (New York: Appleton-Century-Crofts, 1966), pp. 107–169; G.E.W. Wolstenholme, and Maeve O'Connor, eds., *Ethics in Medical Progress With Special Reference to Transplantation.* A Ciba Foundation volume (Boston, Mass.: Little, Brown & Co., 1966); Paul Ramsey, *The Patient as Person: Explorations in Medical Ethics* (New Haven, Conn.: Yale University Press, 1972); Louis Lasagna, *Life, Death and the Doctor* (New York: Alfred A. Knopf, 1968).

21. For example, *see* O. Ruth Russell, "The Right to Choose Death," *New York Times,* February 14, 1972; Robert S. Morison, "Death: Process or Event?" *Science* 173 (August 1971), pp. 694–698; Leon R. Kass, "Death as an Event: A Commentary on Robert Morison," *Science* 173 (August 1971), pp. 698–702; David Hendin, *Death As a Fact of Life* (New York: Norton, 1973).

22. Bernard Barber, *Science and the Social Order* (New York: Collier Books, 1970 [orig. pub. 1952], p. 299.

23. Anon. M.D., *Confessions of a Gynecologist,* p. 75.

24. *New York Times,* September 26, 1970.

25. *Ibid.,* September 29, 1970.

26. *See* Cooper, "Creative Pluralism, Medical Ombudsman," pp. 46–66; Bessman and Swazey, "PKU—A Study of Biomedical Legislation," pp. 49–76; "After Ten Years of PKU Testing," pp. 43–44.

27. Peter Steinfels, "Is Science Stoppable?" *Commonweal* 8, October 2, 1970.

28. For further information *see* "Should Science Have a Conscience?" *Rockefeller Foundation Illustrated* 1 (October 1972), pp. 4–5.

CHAPTER TWO

1. James D. Watson, "Moving Toward the Clonal Man—Is This What We Want?" *Congressional Record-Senate*, April 29, 1971, pp. 12751–12752; "Clonal Propagation," *Genetic Engineering—Evolution of a Technological Issue*. Report to the Subcommittee on Science, Research, and Development of the Committee on Science and Astronautics, U.S. House of Representatives, 92nd Congress, 2nd Sess., November 1972, pp. 21–22.

2. "Genetic Engineering in Man: Ethical Considerations." Editorial in *Journal of the American Medical Association* 220 (May 1, 1972), p. 721. For more popular reports *see*, for example, "Invit: The View from the Glass Oviduct," *Saturday Review*, September 30, 1972; Walter Sullivan, "Implant of Human Embryo Appears Near," *New York Times*, October 29, 1970.

3. Watson, "Moving Toward the Clonal Man," p. 12752.

4. "Invit."

5. Leon R. Kass, "Making Babies—the New Biology and the 'Old' Morality," *The Public Interest* 26 (Winter 1972), pp. 18–56.

6. R. G. Edwards and Ruth E. Fowler. "Human Embryos in the Lab," *Scientific American* 223 (December 1970), pp. 44–54.

7. At this point, Austin referred here to Dr. Patrick Steptoe, the surgeon of Dr. Edwards' team who, together with Dr. Edwards, developed this new technique for extracting eggs from women.

8. Kass quotes Dr. Donald Gould, editor of the British journal, *The New Scientist*, who asked: "What happens to the embryos which are discarded at the end of the day—washed down the sink?" Kass, "Making Babies," p. 32.

9. Paul Ramsey, "Shall We 'Reproduce'? I. The Medical Ethics of *In Vitro* Fertilization," *Journal of the American Medical Association* 220 (June 5, 1972), p. 1347.

10. *Ibid.*

11. For a discussion of the reliability and limitations of various tests, *see* Jay Katz, ed., *Experimentation with Human Beings* (New York: Russell Sage Foundation, 1972), *passim*, specifically, Renee C. Fox, "Experiment Perilous," pp. 369–376; W. St. C. Symmers, Sr., "Not Allowed to Die," p. 709; Joseph D. Cooper, "Creative Pluralism—Medical Ombudsman," pp. 986–992. *See also* Maureen Harris, ed., *Early Diagnosis of Human Genetic Defects*, Fogarty International Center Proceedings 6, HEW Publications 1970, (NIH), pp. 72–75.

12. Kass, "Making Babies," p. 28.
13. Work has been done on other mammals, however, M. C. Chang, of the Worcester Foundation for Experimental Biology in Massachusetts, has done work with rabbits; Joseph C. Daniel, Jr., Professor of Biology at the University of Colorado, used rabbits, ferrets, and minks to study the development of the embryo just before it implants in the uterus; Dr. Wesley Whitten, of the Jackson Laboratory at Bar Harbor, and Dr. John Biggers, of Johns Hopkins, worked with mice eggs *in vitro*. For further details, *see* Edward Grossman, "The Obsolescent Mother," *Atlantic* (May 1971), pp. 39–50.

    Dr. Luigi Mastroianni, is reported to have said: "It is my feeling that we must be very sure we are able to produce normal young by this method in monkeys before we have the temerity to move ahead in the human. . . . In our laboratory, our position is, 'Let's explore the thing thoroughly in monkeys and establish the risk.' " Quoted in Victor Cohn, "Lab Growth of Human Embryo Raises Doubt of 'Normality,' " *Washington Post,* March 21, 1971.
14. Quoted in Grossman, "The Obsolescent Mother," p. 45.
15. *See* Anthony Shaw, " 'Doctor, Do We Have a Choice?' " *New York Times Magazine,* January 30, 1972, pp. 44, 52, 54.
16. *See* Ramsey, "Shall We 'Reproduce'?" pp. 1346–1350. Ramsey reports: "Dr. Patrick Steptoe, Dr. Edwards's colleague, is reported to have said that the decision to implant a given embryo, based on statistical evidence and hope—an embryo which cannot be karyotyped for genetic or other damage as a final procedure before implantation without too grave risk of further, more serious damage—will 'call for a *brave decision.*' " Ramsey goes on to say, "If (as I believe), we should watch our language as we watch our morals, Dr. Steptoe seriously misused language. What he meant was 'rashness' in action, regardless of the consequences to another human life, and not 'courage,' facing one's own perils or adversities," p. 1347.
17. Norman Podhoretz, "Beyond ZPG," *Commentary* 53, 5 (May 1972), p. 6.
18. *Ibid.,* p. 7.
19. Grossman, "The Obsolescent Mother," p. 46.
20. Watson, "Moving Toward the Clonal Man," p. 12751.
21. Alvin Toffler, *Future Shock* (New York: Random House, 1970), p. 177.
22. *Ibid.,* p. 177.
23. "Livestock: Sci-Fi on the Range," *Newsweek,* September 4, 1972.
24. Jane E. Brody, "500 in the U.S. Change Sex in Six Years with Surgery," *New York Times,* November 20, 1972.
25. Kass, "Making Babies," p. 30.

26. *Ibid.*, p. 31.
27. *Ibid.*
28. For a thorough review of experimentation on people *see* Katz, *Experimentation with Human Beings. See also* Bernard Barber, John J. Lally, Julia L. Makarushka, and Daniel Sullivan, *Research on Human Subjects* (New York: Russell Sage Foundation, 1973).
29. Kass, "Making Babies," p. 19.
30. *New York Post*, November 22, 1972.
31. A population's *Attitudes Study* commissioned by WNBC-TV, May 1972, p. 7.
32. Robert G. Edwards and David J. Sharpe, "Social Values and Research in Human Embryology," *Nature* 231 (May 14, 1971), p. 87.
33. For example, *see* Edward Shils, "The Sanctity of Life," *Encounter* 28 (January 1967), pp. 39–49.
34. Kass, "Making Babies," p. 52.
35. Paul Ramsey, *Fabricated Man: The Ethics of Genetic Control* (New Haven, Conn.: Yale University Press, 1970), pp. 1–60.
36. Kass, "Making Babies," p. 50.
37. Herman J. Muller, "Human Evolution by Voluntary Choice of Germ Plasm," *Science* 134 (September 8, 1961), p. 646.

CHAPTER THREE

1. Julian Huxley, "The Future of Man—Evolutionary Aspects," in Gordon Wolstenholme, ed., *Man and His Future* (Boston, Mass.: Little, Brown & Co., 1963), p. 17.
2. Bentley Glass, "Human Heredity and Ethical Problems," *Perspectives in Biology and Medicine* 15 (Winter 1972), p. 243.
3. An often-cited study by Dr. J. A. Fraser Roberts shows that when the risk is low, most parents ignore it; when it is high, 65 percent decide not to have children; another 15 percent find it difficult to make up their mind; only 20 percent go ahead. *See British Medical Journal* 1 (1962), p. 587. *See also* Claire Leonard, Gary Chase, and Barton Childs, "Genetic Counseling: A Consumer's View," *New England Journal of Medicine* 287 (August 31, 1972), pp. 433–439.

CHAPTER FOUR

1. On these terms and their diverse definitions *see* Edward Tatum, "Manipulating Genetic Change," in John Roslansky, ed., *Genetics and the Future of Man* (New York: Appleton-Century-Crofts, 1966), pp. 49–60; Joshua Lederberg, "Experimental Genetics and Human Evolution," *Bulletin of the Atomic Scientists* 23, 8 (October 1966), pp. 4–11; Irene Taviss, "Problems in the Social Control of Biomedical Science and Technology," in Everett Mendelson, Judith P. Swazey, and Irene Taviss, eds., *Human Aspects of Biomedical Innovation* (Cambridge, Mass.: Harvard University Press, 1971), p. 28; Kurt Hirschhorn, "On Redoing Man," *Commonweal* (May 17, 1968), pp. 257–261; "Genetic Engineering: Evolution of a Technical Issue," p. 16.

2. Richard Juberg, "Heredity Counseling," *Nurse's Outlook* 14 (January 1966), pp. 28–33; D. Mechanic, *Medical Sociology: A Selective View* (New York: Free Press, 1968); Jacob J. Feldman, *The Dissemination of Health Information* (Chicago, Ill.: Aldine, 1966).

3. Robert Crain, Elihu Katz, and Donald Rosenthal, *The Politics of Community Conflict* (New York: Bobbs-Merrill, 1969), p. 8; C. S. Rhyne and E. F. Mullin, *Fluoridation of Municipal Water Supply—A Review of the Scientific and Legal Aspects* (Washington, D.C.: National Institute of Municipal Law Officers, 1952), Report No. 140; H. W. Butler, "Legal Aspects of Fluoridating Community Water Supplies," *Journal of the American Dental Association* 65 (November 1962), pp. 653–658.

4. Thomas McCorkle, *Social Science Knowledge of Fluoridation Controversies.* Pennsylvania Papers in Social Science for Public Health Workers 1 (January 1963), p. 2; Crain, Katz, and Rosenthal, *The Politics of Community Conflict*, pp. 243–244.

5. "Genetic Counseling—A Conversation With Dr. Harvey Bender" *a.d. Correspondence* 6, 3 (February 3, 1973).

6. William T. Vukovich, "The Dawning of the Brave New World—Legal, Ethical, and Social Issues of Eugenics," *University of Illinois Law Forum*, 1971 edition, p. 189.

7. Gerald Leach, *The Biocrats*, rev. ed. (Baltimore, Md.: Penguin Books, 1972), p. 137; *Genetic Engineering: Evolution of a Technological Issue*, pp. 34–35.

8. M. Woodside, *Sterilization in North Carolina* (London: Geoffrey Cumberlege, 1950).

9. "The Law: IQ Test," *Newsweek*, October 30, 1972.

10. Mario Biaggi, "Retarded Forgotten," *New York Times*, December 19, 1972.

11. Leach, *The Biocrats*, pp. 139–141.

12. "Population Policy," in Harwood Childs, trans., *The Nazi Primer: Official Handbook for Schooling the Hitler Youth* (New York: Harper & Brothers, 1938).

13. Bentley Glass, *Science and Liberal Education* (Baton Rouge: Louisiana State University Press, 1959), p. 51.

14. H. J. Muller in H. Hoagland and R. W. Burhoe, eds., *Evolution and Man's Progress* (New York: Columbia University Press, 1962), p. 35.

15. John B. Watson, *Behaviorism*, rev. ed. (Chicago: University of Chicago Press, 1962), p. 104.

16. Amitai Etzioni, "Human Beings Are Not Very Easy to Change, After All," *Saturday Review*, June 3, 1972, pp. 45–47.

17. James F. Danielli, "Industry, Society, and Genetic Engineering," *The Hastings Center Report* 2, 6 (December 1972), p. 7.

18. Robert Ettinger, *Man into Superman* (New York: St. Martin's Press, 1972), pp. 54–67.

19. Leach, *The Biocrats*, p. 124.

20. *Ibid.*

21. ". . . no one doubts that a modern slave state would reinforce its class stratification by genetic controls. But it could not do so without having instituted slavery in the first place. . . . It is indeed true that I might fear the control of my behavior through electrical impulses directed into my brain, but . . . I do not accept the implantation of the electrodes except at the point of a gun: the gun is the problem." From Joshua Lederberg, "Genetic Engineering, or the Amelioration of Genetic Defect," *The Pharos of Alpha Omega Alpha* 34 (January 1971), p. 10.

22. Leach, *The Biocrats*, pp. 96–99, 102–104.

23. M. Fox, *Canine Behavior* (Springfield, Ill.: Charles C Thomas, 1965), p. 61.

24. Anon. M.D., *Confessions of a Gynecologist*, p. 74; and private communication.

25. "An Abuse of Prenatal Diagnosis," Letter to the Editor, *Journal of the American Medical Association* 221, 4 (July 24, 1972), p. 408.

26. P. E. Lindahl, *Nature* 181 (1958), p. 784; V. N. Schröder and N. K. Koltsov, *Nature* 131 (1933), p. 329; M. J. Gordon, *Scientific American* 199 (1958), pp. 87–94.

27. F. Unterberger, *Deutsche Medizinische Wochenschrift* 56 (1931), p. 304.

28. Leslie Aldrich Westoff and Charles F. Westoff, *From Now to Zero* (Boston, Mass.: Little, Brown & Co., 1971), p. 122.

29. At a later date (1966), John Peel found in one northern city a continued preference for boys; out of 912 children planned by his sample, 491 desired would be boys; 423, girls. While this makes for a somewhat lower boy surplus than I projected on the basis of earlier data, my data was natural, whereas this data is based on an atypical liberal city. *See* John Peel, "The Hull Family Survey," *Journal of Biosociological Science* 2 (1970), pp. 55–56.

30. See, for instance, Anon. M.D., *Confessions of a Gynecologist*, p. 156. *See also* Leach, *The Biocrats*, p. 87.

31. S. J. Behrman, "Artificial Insemination" in S. J. Behrman and Robert W. Kistner, eds., *Progress in Infertility* (Boston, Mass.: Little, Brown & Co., 1967).

32. "Sperm Banks Multiply as Vasectomies Gain Popularity," *Science* 176 (April 7, 1972), p. 32.

33. Louis Harris, "The Life Poll," *Life,* June 13, 1969, p. 52.

34. *Ibid.*

35. *Ibid.*

36. Congressional Record 117, no. 42 (March 24, 1971), Senate, 1st Sess., p. 1. *See also* Sissela Bok, "The Leading Edge of the Wedge," *The Hastings Center Report* No. 3 (December 1971), pp. 9–11; Paul Ramsey, "The Wedge: Not So Simple," pp. 11–12. *See also* S. E. Luria, "Slippery When Wet." Proceedings of the American Philosophical Society, vol. 116, No. 5 (October 1972), pp. 351–356.

37. Virginia Apgar and Joan Beck, *Is My Baby All Right?* (New York: Trident Press, 1972), p. 172.

38. Lederberg, "Genetic Engineering, or the Amelioration of the Genetic Defect," p. 9.

39. *Ibid.*

40. "Man's Humanity," *Saturday Review of Science* (February 1973), p. 48.

41. Charles F. Westoff and Larry Bumpass, "The Revolution in Birth Control Practices of U.S. Roman Catholics," *Science* 179 (January 5, 1973), pp. 41–44.

42. Westoff and Westoff, *From Now to Zero,* p. 336.

43. James R. Sorenson, "Research Proposal—Genetic Counselors: Professionals in Applied Human Genetics" (Princeton, N.J.: Princeton University, 1971), p. 6.

44. Leonard, Chase, and Childs, "Genetic Counseling: A Consumers' View," p. 433.

45. Sorenson, "Research Proposal—Genetic Counselors," p. 7.

46. "Sickle-Cell Anemia: National Program Raises Problems as Well as Hopes," *Science* 178 (October 20, 1972), pp. 283–286.

47. "Sickle-Cell Anemia: The Route From Obscurity to Prominence," *Science* 178 (October 13, 1972), p. 140.
48. *Ibid.*
49. *Ibid.,* p. 141.
50. "Cooley's Anemia: Special Treatment for Another Ethnic Disease," *Science* 178 (November 10, 1972), p. 592.
51. *Ibid.,* pp. 590–593.
52. *Ibid.*

CHAPTER FIVE

1. Barber, Lally, Makarushka, and Sullivan, *Research On Human Subjects.*
2. Jean Heller, "Syphilis Victims in U.S. Study Went Untreated for 40 Years," *New York Times,* July 26, 1972.
3. "Kennedy Says 45 Babies Died in a Test," *New York Times,* October 12, 1972.
4. *Ibid.*
5. Sir John Eccles, "Experiments on Man in Neurophysiology," in V. Fattorusso, ed., *Biomedical Science and the Dilemma of Human Experimentation* (Paris: Council for International Organizations of Medical Sciences, 1967), pp. 22–23.
6. H. F. Wiese, A. E. Hanson, and D. Adam, "Essential Fatty Acids in Infant Nutrition," Parts I, II and III, *Journal of Nutrition* 66 (1958).
7. *Wall Street Journal,* February 22, 1973.
8. For discussion of the issues related to the definition of death, *see* Morison, "Death: Process or Event?" pp. 694–698; Kass, "Death as an Event: A Commentary on Robert Morison," pp. 698–702; David White, "Death Control," *New Society* (November 30, 1972), pp. 502–505; Warren T. Reich, "Dignity in Death and Life," *New York Times,* January 16, 1973; O. Ruth Russell, "The Right to Choose Death," *New York Times,* February 14, 1972; Hendin, *Death as a Fact of Life.*
9. For further discussion, *see* David Daube, "Transplantation—Acceptability of Procedures and the Required Legal Sanctions," in *Ethics in Medical Progress—With Special Reference to Transplantation* (Boston, Mass.: Little, Brown & Co., 1966), pp. 198–199.
10. R. A. Spitz, "Hospitalism: An Inquiry into the Genesis of Psychiatric

Conditions in Early Childhood," *Psychoanalytic Study of the Child* 1 (1945), pp. 53–74; R. A. Spitz, "Hospitalism: A Follow-up Report on Investigation Described in Volume 1, 1945," *Psychoanalytic Study of the Child* 2 (1946), pp. 113–117. Numerous experimental studies on animals have been done in this area; *see,* for example, S. Svomi, "Repetitive Peer Separation of Young Monkeys," *Journal of Abnormal Psychology* 81, 1 (February 1973), pp. 1–10.

11. Oliver Cope, "Breast Cancer—Has the Time Come for a Less Mutilating Treatment?" *Radcliffe Quarterly* 54, 4 (1970), pp. 6–11; "The Breast—Cancer Debate," *Newsweek,* November 16, 1970, p. 121; Eileen Shanahan, "Women Unionists Back Equal Rights Plan; Doctor Alleges Unnecessary Surgery by Men," *New York Times,* September 15, 1970; Jane E. Brody, "New Cancer Treatment View Scored," *New York Times,* May 20, 1971; S. Kaae and H. Johansen, "Breast Cancer; Five-Year Results: Two Random Series of Simple Mastectomy with Post-Operative Irradiation versus Extended Radical Mastectomy," *American Journal of Roentgenology, Radium Therapy, and Nuclear Medicine* 87 (1962), pp. 82–88; B. Fisher, "The Surgical Dilemma in the Primary Therapy of Invasive Breast Cancer: A Critical Appraisal," *Current Problems in Surgery* (October 1970), pp. 3–53.

12. For a discussion of this issue, *see* Dr. George Crile, Jr., "Breast Cancer and Informed Consent," *Chicago Clinic Quarterly* 39, 2 (Summer 1972), pp. 57–59. *See also* his book *A Biological Consideration of Treatment of Breast Cancer* (Springfield, Ill.: Charles C Thomas, 1967).

13. Crile, "Breast Cancer and Informed Consent," p. 59.

14. Jerry Bishop, "The Heart Transplant Advances at Stanford but Halts Elsewhere," *Wall Street Journal,* December 4, 1972; Harold Schmeck, "Experts Say Human Organ Transplants Are at Level Below What Doctors Would Like," *New York Times,* September 26, 1972.

15. Bishop, "The Heart Transplant Advances at Stanford but Halts Elsewhere"; Schmeck, "Experts Say Human Organs Are at Level Below What Doctors Would Like"; "Kidney Transplants: How to Save Life With a Registry," *New York Times,* September 19, 1971.

16. John A. Clausen, "Mental Disorders," in Robert K. Merton and Robert Nisbet, eds., *Contemporary Social Problems* (New York: Harcourt Brace & World, 1961), p. 137.

17. H. Basowitz, H. Persky, S. Korchin, and R. Grinker, *Anxiety and Stress* (New York: McGraw-Hill, 1955), pp. 12–13; *see also* Hans Selye, *The Stress of Life* (New York: McGraw-Hill, 1956).

18. *See* Barbara Seaman, *Free and Female* (New York: Coward, McCann & Geoghegan, 1972), pp. 140–147; Willard Gaylin, "The Patient's Bill of Rights," editorial, *Saturday Review of Science* (March 1973), p. 22.

19. Edward M. Kennedy, *In Critical Condition—The Crisis in America's Health Care* (New York: Simon & Schuster, 1972).

20. Zena Stein, Mervyn Susser, and Andrea V. Guterman, "Screening Programme for Prevention of Down's Syndrome," *The Lancet* (February 10, 1973), pp. 305–310. *See* several reports, not all as optimistic, in Maureen Harris, ed., *Early Diagnosis of Human Genetic Defects* (Washington, D.C.: U.S. Government Printing Office, 1970).

21. Private communication.

22. *See* Barbara Seaman, "Do Gynecologists Exploit Their Patients?" *New York Magazine* (August 14, 1972), pp. 47–54; "Vaginal Infections," *Up From Under* 1 (Winter 1971–72), pp. 23–28; Seaman, *Free and Female*, pp. 139–182; "The Yogurt Cure," *Newsweek,* December 18, 1972; Ellen Frankfort, *Vaginal Politics* (New York: Quadrangle, 1972); Susan Bondurant, "It's All Right, Doc—I'm Only Dying," *Rough Times* (West Somerville, Mass.: The Radical Therapist, Inc., 1972), p. 10; "The Male-Feasance of Health," *Health-Pac* (New York: Health Policy Advisory Center, Inc., March 1970); Eileen Shanahan, "Women Unionists Back Equal Rights Plan: Doctor Alleges Unnecessary Surgery by Men," *New York Times,* September 15, 1970; Glenda Adams, "Natural Childbirth: Just Another Shuck," *Village Voice* (September 30, 1970); "Surviving the Hospital," *Rough Time* (West Somerville, Mass.: The Radical Therapist, Inc., 1972); "Women, Medicine, and Capitalism," *Our Bodies, Our Selves* (Boston Women's Health Course Collective, 1971), pp. 134–136; Betty Friedan, "Up From the Kitchen Floor," *New York Times Magazine* (March 4, 1973), pp. 8–9; Diana Scully and Pauline Bart, "A Funny Thing Happened on the Way to the Orifice: Women in Gynecology Textbooks," AJS, 78, 4, 1045–1050.

23. Anon. M.D., *Confessions of a Gynecologist,* p. 157; Leach, *The Biocrats,* p. 87.

24. "Children of Incest," *Newsweek,* October 9, 1972. Copyright Newsweek, Inc., 1972, reprinted by permission.

25. For a mother's report that "the neurosurgeon made the decision" that her deformed newborn will be operated on, *see* Luree Miller, letter to the Editor, *Commentary* (October 1972), p. 28.

26. Seaman, *Free and Female*, pp. 140–147; Gaylin, "The Patient's Bill of Rights."

CHAPTER SIX

1. The data are from Charles F. Westoff, "Modernization of U.S. Contraceptive Practice," *Family Planning Perspectives* 4, 3 (July 1972), p. 10.
2. Westoff and Westoff, *From Now to Zero*, pp. 53–54. *See also* Christopher Tietze, "Intra-Uterine Conception," in Stewart Marcus and Cyril Marcus, eds., *Advances in Obstetrics and Gynecology* 1 (1967), pp. 486–593.
3. Westoff and Westoff, *From Now to Zero*, pp. 324–330.
4. Robert G. Potter and James M. Sakoda, "A Computer Model of Family Building Based on Expected Values," *Demography* 3 (1966), pp. 450–461.
5. W. H. W. Inman and M. P. Vessey, "Investigation of Deaths from Pulmonary, Coronary, and Cerebral Thrombosis and Embolism in Women of Child-bearing Age," *British Medical Journal* 2 (April 27, 1968), p. 199.
6. M. P. Vessey and Richard Doll, "Investigation of Relation Between Use of Oral Contraceptives and Thromboembolic Disease," *British Medical Journal* 2 (April 27, 1968), p. 203.
7. Alan F. Guttmacher, *Pregnancy, Birth, and Family Planning* (New York: Viking Press, 1973), pp. 277–278.
8. Westoff and Westoff, *From Now to Zero*, p. 92.
9. *Ibid.*, pp. 86–116.
10. *Ibid.*, p. 94.
11. *Ibid.*, p. 97.
12. *Ibid.*, pp. 96–99.
13. *Ibid.*, p. 99; Seaman, *Free and Female*, pp. 232, 229–231.
14. Westoff and Westoff, *From Now to Zero*, p. 100; Seaman, *Free and Female*, p. 232.
15. Seaman, *Free and Female*, p. 232.
16. *Ibid.*, p. 232.
17. Westoff and Westoff, *From Now to Zero*, p. 52.
18. *Ibid.*, p. 101.
19. *Ibid.*, pp. 101–102.
20. Seaman, *Free and Female*, p. 221.
21. Leach, *The Biocrats*, p. 29.
22. Arthur Frank, M.D. and Stuart Frank, M.D., *The People's Handbook of Medical Care* (New York: Random House, 1972), p. 305.
23. On differential utilization, *see* Westoff, "The Modernization of U.S. Contraceptive Practice," p. 12.
24. Westoff and Westoff, *From Now to Zero*, p. 99.

25. Morton Mintz, "Drug Official Cited Danger of Pill in 1971," *Washington Post,* December 10, 1972.
26. Leach, *The Biocrats,* p. 29.
27. Westoff and Westoff, *From Now to Zero,* p. 99.
28. For more on this, *see* Salhanick's book, *Metabolic Effects of Gonadal Hormones and Contraceptive Steroids* (New York: Plenum Press, 1969).
29. "Birth Control: Current Technology, Future Prospects," *Science* 179 (March 23, 1973), p. 1222.
30. *Ibid.*
31. G. Barsy and J. Sarkany, *Journal Demografia* 6 (1963), pp. 427–467; K. Miltenyi, *Journal Demografia* 7 (1965), pp. 73–87; Y. Moriyama et al., "Harmful Effects of Induced Abortion," *Family Planning* (Tokyo: Family Planning Federation of Japan, Subcommittee on the Study of Induced Abortion, 1966), pp. 64–73; Charles Wright, Stuart Campbell and John Beazley, "Second-Trimester Abortion After Vaginal Termination of Pregnancy," *Lancet* 1 (June 10, 1972), pp. 1278–1279.
32. A. Kotasek, *International Journal of Gynecology and Obstetrics* 9 (1971), p. 118.

CHAPTER SEVEN

1. Ivan Illich, "Planned Poverty: The End Result of Technical Assistance," in *Celebration of Awareness* (New York: Doubleday, 1970), p. 162.
2. *Ibid.,* p. 163.
3. Jane E. Brody and Edward B. Fiske, "Ethics Debate Set Off by Life Science Gains," *New York Times,* March 28, 1971.
4. *Ibid.*
5. These are the figures of Dr. Vernal Cave, reported in Philip H. Dougherty, "Advertising: City to Begin VD Campaign," *New York Times,* January 28, 1972.
6. Unpublished data from the National Center for Health Statistics, reported in Harold Schmeck, "Research Funds and Disease Effects Held Out of Step," *New York Times,* February 10, 1973.
7. John J. Fried, "The Bloody Pressure on 22 Million Americans," *New York Times Magazine,* February 25, 1973.
8. Edward Kennedy, *In Critical Condition—The Crisis in America's Health Care* (New York: Simon & Schuster, 1972).
9. *Ibid.,* p. 155.

10. *Ibid.*, p. 162.
11. *Ibid.*, pp. 13, 15.
12. Barber, *Science and the Social Order*, p. 299.

POSTSCRIPT

1. Robert Rosenthal, *Experimenter Effects in Behavioral Research* (New York: Appleton-Century-Crofts, 1966), p. 161. For more on the consequences of labeling, *see* Amitai Etzioni, "The Stigma of Names and the Case of Local TV Programming Seen in Perspective: Thou Shalt Not Label," *Human Behavior* 1 (September/October, 1972), pp. 6–7.
2. Rosenthal, *Experimenter Effects in Behavioral Research*, p. 411.
3. *Journal of the American Medical Association* 222, 2 (October 9, 1972); *See also* Burke, "The XYY Syndrome," pp. 261–284; Telfer et al., "Incidence of Gross Chromosomal Errors," pp. 1249–1250.
4. "Sickle-Cell Anemia: The Route From Obscurity to Prominence," p. 138.
5. *Ibid.*, pp. 138–139.
6. V. Cohn, "Disease Publicity Raises Problems," *Washington Post*, November 12, 1972; "Sickle-Cell Anemia: The Route From Obscurity to Prominence," p. 141.
7. Cohn, "Disease Publicity Raises Problems"; "Sickle-Cell Anemia: The Route From Obscurity to Prominence," p. 141.
8. Jane E. Brody, "Problems Seen in Genetics Tests," *New York Times*, March 25, 1972.
9. *Newsweek*, March 26, 1973; Apgar and Beck, *Is My Baby All Right?*, pp. 206–207.
10. It is only lately that one airline—United—has decided that having the sickle-cell *trait*, though not the disease, will not disqualify blacks from being employed as cabin attendants. *Wall Street Journal*, March 27, 1973.
11. Gaylin, "The Patient's Bill of Rights," p. 22.
12. News release of the Medical Society of the State of New York (January 11, 1973).
13. *Ibid.*
14. Jane E. Brody, "Hospitals Prepare for 1.6 Million Abortions Annually," *New York Times* (January 28, 1973).
15. *Ibid.*
16. *See* Yukio Manabe, "Danger of Hypertonic-Saline-Induced Abor-

tion," Letter to the Editor, *Journal of the American Medical Association* 210, 11 (December 15, 1969), p. 2091.

17. Brody, "Hospitals Prepare for 1.6 Million Abortions Annually."
18. See Laurie Johnston, "Abortion Clinics in City Face Rising Competition," *New York Times,* March 19, 1973.
19. Bernard Barber, *The Problems of Human Experimentation.* Hearings before the Senate Sub-Committee on Health of the Senate Committee on Labor and Public Welfare, 93rd Congress, March 8, 1973.
20. Jay Katz, Opening Statement. Hearings before the Senate Sub-Committee on Health of the Senate Committee on Labor and Public Welfare, 93rd Congress, March 8, 1973.

Indexes

Name Index

271

Subject Index